DAVID McKIE

KW-484-054

A SADLY MISMANAGED AFFAIR

A Political History of the Third London Airport

CROOM HELM LONDON

FIRST PUBLISHED 1973
© 1973 BY DAVID MCKIE

CROOM HELM LTD, 2–10 ST JOHNS ROAD LONDON SW11

ISBN 0–85664–096–4

KIRKCALDY PUBLIC LIBRARIES

703/77

629.36 MAC

KIRKCALDY

DISCARD
10/85

DISTRICT LIBRARIES

PRINTED IN GREAT BRITAIN
BY EBENEZER BAYLIS & SON LTD
THE TRINITY PRESS, WORCESTER, AND LONDON
BOUND BY G. & J. KITCAT LTD, LONDON

A Sadly Mismanaged Affair

Contents

The need now is for an urgent inquiry into the whole process of decision-making in government. Stansted is not the only ex-example – even if it is the most flagrant – of the system going wrong, and we cannot possibly afford any further errors on this scale. Surely it is now the duty of the Government, in the light of these events, to set up some kind of inquiry to see how decisions of this kind are made – and how they can be prevented in future.

Peter Kirk, Conservative MP for Saffron Walden, in a letter to *The Times*, 5 March 1969.

What lessons can we learn from this (as I freely admit it to be) sadly mismanaged affair?

Lord Walston, former Parliamentary Secretary at the Board of Trade, in the Lords debate on Stansted, 11 December 1967.

We have to plan far ahead, but we must learn from the example of Thales, who, while looking at the stars, fell into a well.

Ivor Thomas, Parliamentary Secretary to the Ministry of Civil Aviation, in a Commons debate, 24 January 1946.

TO MY MOTHER

1

In which a judge is sent for; and new peaks of rationality are scaled

At about three o'clock on the afternoon of Wednesday, 12 August 1970, the honourable Mr Justice Roskill, judge of the High Court, leaned back in his chair in a banqueting room in the Piccadilly Hotel in London, lit a cigarette and (according to the next day's *Daily Telegraph*) said: 'Now I understand they will be having this room back for bingo and bridge.'

Sir Eustace Roskill and his six colleagues on the Commission on the third London airport had just completed what at that time was the longest public planning inquiry in the country's history. Its seventy-four days had, admittedly, failed quite to equal the seventy-six which Roskill had spent in his more accustomed role on the case of the stranding of the ship *Medina Princess* at Djibouti; but in the less glamorous and protracted world of planning debate, the Roskill operation had lasted longer, commissioned more research, staged more public argument, involved more people (from lawyers to ordinary men in the street) and, at £1·2 million, cost more than anything before it. The awed murmurs of those who said there had been nothing like it before quite drowned the muttered determination of others that there would be nothing like it again.

Against the evidence of 160 witnesses, many of them great with authoritative submission, the Piccadilly inquiry had been able to set a bank of independent research which no previous planning operation could have matched. For Roskill had been armed with a research team, twenty-three strong, at home in transport, civil engineering, planning control and statistical analysis, which any civil service department, embarked on a major planning project, might have given its eyes to get. The research team, what was more, was equipped to construct a cost benefit analysis, a system devised in the United States, which had already proved its worth by justifying the addition to the London Underground of the Victoria Line, a scheme which many agreed was desirable but which it was somewhat more difficult to justify in conventional costing terms.

'The best available aid to rational decision-making' they called it in their report. It was a step beyond the crude and largely subjective procedures in which you simply made a straight assessment of the costs of various options and then decided which would bring the best return. Cost benefit analysis could search out and evaluate factors which a simpler analysis would altogether miss. It would spot, and rope into its calculations, side-effects on people who would never come near the airport and on places hundreds of miles away.

It was Roskill's task to construct an edifice of pure logic in which the subjective would (as far as was humanly possible) be banished, and the objective would be given its rightful place. As justice, traditionally, weighs every scintilla of evidence, on either side, and produces a final result with the objectivity of one who is blindfold, so the Roskill team would produce a final report in which every consideration of airport selection and construction, from local pride to the supply of aggregates, would find its proper place and weight. This concept of balancing was at the centre of the Commission's operation; and it was because many who pronounced on its findings had failed to appreciate the subtlety of the balancing on which they were based that so much of the subsequent comment did Roskill and his colleagues such injustice.

They were to say, in setting out their findings:

We believe therefore that, following our refusal to accept the existence of any absolute constraint upon the choice of site, the right answer in the interests of the nation rests in a choice of site which, however damaging to some, affords on a balanced judgment of advantages and disadvantages the best opportunity of benefiting the nation as a whole. [And again:] The painful and difficult choices which have to be made arise from the attempt in a democratic society to find an acceptable balance between conflicting interests. We have viewed our work as part of that process.

People talked as if one could simply weigh the sufferings of the people of Cublington, should the airport be inflicted upon them, with the sufferings of a much smaller group of people, should that be the choice, at Foulness. Not so, said Roskill. If the airport went to Foulness, the attractions of Heathrow, of Gatwick, of Luton would be increased; the noise inflicted on the populations there would then increase also. Their interests must be taken into the great equation, too. Again, the interests of those who used the airport must be balanced against those of the populations on whom it was visited. There was a danger, said Roskill, and it was on this that his conclusion rested, that the airport which caused least local suffering might be one which, in terms of airport business, did not adequately work. No solution could be reached without loss. Lose *these* good things (the clustering villages, the fine churches, the solid comfortable communities of north Buckinghamshire) and get an airport that is of value; or lose these *other* things, also valuable – the remote homesteads of farthest Essex, the Brent geese on the estuary – and get an airport which may not work; and which, at the same time, will not relieve the present troubles of communities elsewhere. This, basically, was the kind of balance (though any simplification distorts it, for it is an argument of extraordinarily intricate construction) on which the recommendation of the Roskill Commission to reject the 'popular' solution, Foulness, in favour of the more efficient and balanced solution, Cublington, was based.

All admired the edifice that Roskill had built; many pointed out with quite unfeigned respect the care and expertise with which this monument had been constructed. But few really liked it very much. The master-builder had skilfully set himself to carry out his instructions; no one, it was conceded, could have done it better. But somehow, having been built, it no longer seemed the right thing to build.

But there was also – and this was of the greatest importance, since it gave a much stronger mandate for doubt outside – a doubter, a disaffected member, within the Commission itself. Colin Buchanan, a civil servant of unremarked importance, had been sought out seven years before by Ernest Marples, then Minister of Transport, to examine the problems created for towns and cities by the rapid spread of the motor car. Marples, it was said, had sent for the best books on the subject, and deciding that Buchanan's was the best of all, had ordered that its author be sent for; at which a search was made, and he was found somewhere in the corridors of the Housing Ministry. The report of Buchanan's team, *Traffic in Towns*, had created for Buchanan a reputation both as a champion of the environment and as its destroyer. Much of his report was taken up with the defence of environmental values, a defence which in those days was the prerogative of a few, popular though it has now become. Yet to save our towns from total destruction, much, Buchanan said, would have to be sacrificed. We must build roads on a scale few had contemplated. And it was easy to see from the theoretical schemes contained in the report, for places as diverse as Leeds and Newbury, that this would not be done without cost. 'Buchananism,' said a critic, 'would cost the earth and would apparently remove us from contact with it.'

Buchanan – as he later admitted – was in doubt within a few months of the Commission's launching about the direction its work was taking. When it appeared that, true to the logic of its scientific analyses, the Commission might not find place for Foulness on its short list at all, it seemed likely that Buchanan would resign. In the event, Foulness was included. But Buchanan remained disaffected; and with the end of the Commission's work some three months away, he made it plain that the conclusion to which it was heading was one to which he could not subscribe. That which, to his colleagues, had seemed like the clear indication of quantified common sense seemed to Buchanan almost like madness. They accepted no overriding constraint; he raised one like a banner. It was the environmental aspects of the airport, the planning considerations, which must override the rest. Any of the three inland sites – Cublington in Buckinghamshire, Thurleigh in Bedfordshire, Nuthampstead in Hertfordshire – would be an environmental disaster; but Cublington would be the most disastrous of all.

But there was more to his disagreement than that. The fundamental difference between Buchanan and his colleagues was a philosophical one and one too deep ever to have been resolved.

Roskill stood for rationality. 'The highest monument yet erected to Whitehall rationality', *The Times* called the Commission. 'The grandest exercise in pure reason which any British government has yet mounted', said Hugo Young in the *Sunday Times*. The geographer, Peter Hall (who was to emerge as a critic of the Commission), saluted, 'the most rational dispassionate procedure that good minds could devise'. Indeed, it is possible to see Roskill as the culmination of one of the dominant trends of the political sixties; the conviction that the rational and the efficient, rather than the picturesque and the sentimental, must prevail.

In the late fifties, a succession of books had lectured the British on the need to turn their backs on the past if the ills of the nation, the sluggish performance in world leagues, the laggard rate of economic growth, were ever to be corrected. It was tempting, they said, to wallow in the warm water of nostalgia, to look back on past glories, on England as it used to be. But it would be our national undoing. We must turn away from the world of the garden party (and grouse moor), from the romance of the branch line, its bucolic charm exceeded only by its weight of government subsidy. We must learn to understand, and employ, the cost analysis, the computer, the advanced machine.

It was this mood which, at the end of the fifties, set Beeching to rationalise the railways, and McKinsey, it sometimes seemed, to work on almost everything else. It was the harnessing of this mood which helped Harold Wilson capture the middle ground in the elections of 1964 and 1966. As Michael Shanks had written in one of the most characteristic of the What's-Wrong-with-Britain books – *The Stagnant Society* (subtitled: 'A Warning')[1] in 1961 – 'what sort of an island do we want to be?'

> This is the question to which we come back in the end. A lotus island of easy, tolerant ways, bathed in the golden glow of an imperial sunset, shielded from discontent by a thread-bare welfare state and an acceptance of genteel poverty? Or the tough dynamic race we have been in the past, striving always to better ourselves, seeking new worlds to conquer in the place of those we have lost, ready to accept growing pains as the price of growth? . . . At every stage of the journey there will be the temptation to refuse the hurdle, to take the easy way out, to go back to lotus land and impoverished gentility, to turn in on ourselves and try to forget the bustling menacing ungrateful world outside.

Roskill, it seemed, was worried about lotus land too. His medicine might have a bitter taste, but it might still do us more good in the end than other medicines which tasted sweeter. 'The nation's unsatisfactory economic performance in recent years can at least in part be attributed to a national tendency to forgo economic gains and to prefer other goals,' said the majority recommendation. And again: 'The nation cannot afford to decide the site for this airport on the basis of a serious misallocation of scarce resources.' There was a substantial risk that Foulness would long be a liability to the taxpayer; the difficulty of getting there might 'sway the balance in a decision whether or not to make a potentially valuable business trip'. And, in a characteristic passage (though phrased with an awkwardness quite uncharacteristic of the Roskill Report):

> To choose a site which though it may reduce the disadvan-tages also greatly reduces the advantages affords little chance of ultimate benefit to the nation.
> To choose a site which, though it possesses disadvantages,

also yields greater advantages, increases the ultimate gain to the nation, not least because the nation's wealth is thereby increased and through that increase in the nation's wealth can come the greatest opportunity of compensating those who will suffer the disadvantages.

All this was a foreign language to Buchanan. To 'forgo economic gains and to prefer other goals' was not, for him, the irresponsible shortsightedness it seemed to his colleagues, but the mark of maturity and civilised standards.[2]

Against the dedicated rationality of the Roskill majority, he argued his case on boldly non-rational grounds. Against the great acres of quantification he set those considerations which by their essence could never be quantified. He readily admitted that the sophisticated arguments on cost benefit had often been beyond him; but he did not think he had suffered as a result.

> If this opinon [he wrote] implies, as I think it must, that I do not wholly accept the results of the cost benefit analysis, then this is the truth of the matter. The analysis has caused me more worry and heart-searching than anything I have previously encountered in my life. From the start of the Commission's work I have stood in humble awe of the resource and ingenuity of the research team in undertaking the analysis. The intellectual qualities they displayed coupled with their dispassionate approach have made me acutely aware of my own deficiencies. At the same time, paradoxically, I have been beset with anxieties. As the team progressed, with ever more ingenious methods of surmounting this or that difficulty or criticism, so I became more and more anxious lest I be trapped in a process which I did not fully understand and ultimately led without choice to a conclusion which I would know in my heart of hearts I did not agree with.[3]

The campaigners for Cublington had called as one of their witnesses the poet and preservationist, John Betjeman; a man who regarded efficiency as deeply to be suspected, the epitome of every unmodern-minded man who ever became more anguished over a threatened portico than over a further decline in our visible exports. They had been richly rewarded. Arriving at Aylesbury for the local hearings, characteristically contriving to be late, he had announced himself as a man who visited many parts of England 'using my eyes, nose and ears'. The scenery here in north Buckinghamshire, he said, was of that very sort which Henry James had described as 'unmitigated England'. The long street of Stewkley – one of three villages threatened with extinction – was 'a true story of the growth of the century'.

> This is not, I realise, a matter of economics. One cannot assess in terms of cash, or exports or imports, an imponderable thing like the turn of a lane or an inn or a church tower or a familiar sky line. Just try driving as I did with the wind behind you and a western sun on the leaves of the elms and oaks, along the pleasant undulations from Winslow via Swanbourne, Drayton Parslow and Hollingdon to Soulbury and you will see what I mean.

Only the choice of Stansted, said Betjeman, could be worse. And now Buchanan, cutting free from the rationality of his colleagues, also gave evidence for those values which could not be measured. He described his feelings, standing on Ivinghoe Beacon and looking out across the Vale of Aylesbury, that to put an airport in such a place would be unthinkable.

> At Stage Five there was a picture on the wall of the corridor in the Piccadilly Hotel – a reproduction of a picture of the Vale by Rex Whistler . . . It was as potent as any of the evidence I heard . . . the view north-westwards from the Chilterns is as rewarding as any in the south of England . . . the ridge road from Whitchurch through Oving to Pitchcott and Waddesdon is as beautiful a road (for its views) as any I know in the Home Counties.

This curious and unexpected quality was at the heart of Buchanan's submission; it was the values of Betjeman expressed by the author of *Traffic In Towns*.

And just as those now distant condemnations of national inefficiency, of the preference for sentiment over reason, had struck a national chord, so the language of Buchanan did now. For if economic efficiency had been the prevailing political fashion of the start of the decade, the environment was the great new enthusiasm of its end. Labour, coming to power, created two new prestige ministries: Economic Affairs, designed to counterbalance and penetrate the superior mystique of the Treasury, and Technology, designed to rationalise and reshape British industry to do battle with the competitive world outside. The Conservatives created a new Department of the Environment; and there was no party distinction about that, for everyone knew that had Labour won, Anthony Crosland would have taken on the empire which now fell to Peter Walker.

The desirability of enshrining growth as a target above all others, of being ready to forgo other goals for economic gain, was increasingly under challenge. A series of decisions, Buchanan complained, each said to be essential to our economic survival, had in their accumulation begun to produce a country not worth surviving in. Here, he declared, was a test case of the seriousness of our commitment to the values of the environment and of humanity. On 26 April 1971 the Conservative Government accepted his challenge, and announced that the Roskill majority recommendation had been overthrown. 'On environmental and planning grounds,' said the Secretary of State for Trade and Industry, John Davies, 'the Foulness site is the best. And the Government have concluded that these considerations are of paramount importance.'

The Government's choice, it appeared, was philosophical. The strictest demands of economic logic were to be ignored because the Government – a government, moreover, which had come to power with a warmer regard for the laws of the market than any for many years before it – had perceived the existence of a higher good. Yet some of this was packaging. It may well be that the environmental factors would always have weighed heavily; had not the Prime Minister himself declared not long before (though it should be noted that he was speaking to the Countryside Commission) that sometimes the choice which appeared the best economic prospect in the short term should be rejected because of its long-term implications? But in fact, the Government made its choice, as governments usually do, on the basis of the political logic of the situation. That is what governments are there for. The airport had to go to Foulness not simply because of any conscious resolve that environmental arguments should weigh heavier than they formerly did, and economic arguments less heavily; but because the political facts of life decreed that this should be so.

But how – it may be asked – did a government, pledged, as the Labour Government was, to rationality and planning, come to assign such a heavy share of the responsibility for one of the most crucial investment and planning decisions in British history to an outside group – comprising, it is true, an economist, an aviation specialist and a planner, but presided over by a man who could justly be said (indeed, it was his main qualification) to have no expert knowledge of airports at all? And was it ever rational to try to place such an inescapably political matter outside the hands of politicians?

In fact – like the final decision on Foulness – this was a choice more or less forced on Whitehall by accumulated circumstances over which government no longer maintained an unchallenged control. The decision to call in Roskill, to set up a commission which would provide a guarantee of total impartiality, and which (it seems to have been hoped) would produce an answer of unimpeachable *certainty*, instead of the thinly argued and deeply contentious solutions which had been adopted before – this was the natural consequence of the tangled and incongruous history of airport planning since the war.

Both with Stansted and with Gatwick before it, there had been decisions – or sometimes, perhaps more damagingly, non-decisions or assumptions which came to take on the status of decisions in the process of time – which had undermined public confidence in the ability of Whitehall to produce a fair and expert judgment. An atmosphere of protest and grievance infected the whole debate. Roskill would provide a fresh start.

The dream failed, as attempts at the impossible usually fail, because the logic of the economist, the impartial dispensation of the judge, do not always square with the judgments of the politician or the man in the street. Yet the tributes paid to Roskill in Lords and Commons by people who then went on to question the whole basis of his operation were not mere charade. At least the final decision on the third airport could now be taken, as it never could have been during the Stansted affair, on the basis of a comprehensive bank of evidence; at least the risks involved in rejecting the best economic solution could be coolly faced. People said it was ruinously expensive, that it was a farce to spend so much money on an inquiry which produced a solution with so little chance of acceptance. Yet £1·2 million (the cost to the tax-payer – add in what the interested parties paid for their research, their local activity, their legal representation and you could comfortably double that figure) does not seem an excessive cost to pay for an examination as searching as Roskill's. It is an in-finitesimal proportion of the cost of building a modern airport. It is little more than the cost of a mile of motorway. And it arguably saved us from investing hundreds of million pounds in the wrong place at Stansted.

In the end, Roskill's principal contentions were rejected; and that rejection has left some bitterness in some of those who were involved. But it was, even so, a major advance on anything which had been done before. Its merits, ironically, were only to be appreciated fully once the plans to build on Maplin Sands, in disregard of what Roskill recommended, began to take vast and expensive shape two years later.

2

*. . . In which airfields are replaced by airports,
but decisions are made much as they were before*

In pre-war Britain, civil aviation was still largely in a primitive,
if picturesque condition. Many of the airfields were sketchy and
relied heavily on improvisation. In a Commons debate in March
1949 the Member for Caithness, E. L. Gandar Dower, recalled
his own pioneering days an as aeronautical entrepreneur:

> One went to a district, selected a large grass field, per-
> suaded the Ministry of Civil Aviation to give it their blessing
> and accepted responsibility for using the airfield with
> safety. One obtained a car with which to run the passengers
> to the nearest village, and one man acted as telephone
> operator, booking clerk, ticket collector and everything
> else.

The first London airport was on Hounslow Heath. It was
taken from the military authorities to become the first established
London terminal and the first regular daily passenger service in
the world flew out on its inaugural journey on 25 August 1919.
No other country, it was admiringly observed, had an organisation so fully developed for dealing with aeroplanes arriving
from overseas. Indeed, it was only six years before this that the
first piece of aerodrome law on the statute book – the Aerial
Navigation Act of 1913 – had enabled the government to designate areas for landing. Before then, the incoming aviator could
touch down wherever he chose.

Almost immediately, however, there seems to have been a
change of plan. For in 1920, the War Office claimed Hounslow
Heath back again, on the unarguable grounds that it was needed

to train the cavalry. It was a prophetic start to the story of London's airports, which was to be riddled with plans made only to be abandoned – and with frustration of civil activity by the designs of the departments of Defence.

The chosen successor was Croydon, on the southern fringe of London, which was bought for this purpose in 1920. It was easier to get to by road; it was also judged to have better weather. The first simple airfield was improved in 1923, 1925 and again in 1928; by 1939, total capital expenditure on the site was reckoned to have reached £400,000.

The airport was the focus of pride and popular affection. People came for the day to watch the planes landing and taking off; in 1929 the total number of visitors was put at 16,000. By 1937 it was thought necessary to take some of the weight off an increasingly busy Croydon by opening a new Air Ministry civil airport at Heston, not far from Hounslow. But by this time there was growing concern about the unco-ordinated and haphazard growth of civil aviation. A committee was set up under Sir Henry Maybury, formerly Director-General of Roads and consultant adviser to the Ministry of Transport, and the result was a recommendation for a ring of airports around the capital, with Croydon and Heston among them.

There were other candidates for inclusion in the ring. The policy of the Air Ministry was to see airports proliferate provided that this could be done without heavy government involvement. In October 1928, the Ministry wrote to town clerks in all towns of above 20,000 people calling the attention of their corporations to the need to establish municipal civil aerodromes and offering expert advice in choosing sites. (Towns, of 20,000 at that time included such places as Bacup and Glossop, Brighouse and Bexhill.) The response was sluggish. A mere twenty-one called the Minister's inspectors in after that appeal was launched; by 1933, only sixteen municipal aerodromes had been licensed. The Ministry decided that municipal enterprise must be fired by a public display of royal enthusiasm. A conference was convened at the Mansion House in December 1933 at which the principal speaker was the Prince of Wales, who obliged with the assertion that the lack of adequate local authority landing grounds was handicapping the whole development of commercial aviation in Britain. This conference resulted in a new burst of interest, and by the outbreak of war there were forty-two municipal airports.

But most of these municipal airports were unambitious and unattractive. There was not a hard runway among them – nor indeed was there a hard runway at Heston or Croydon. The feeling was growing that Britain was falling badly behind in the civil aviation world. The Air Ministry was coming under sharp attack in Parliament. At the end of a heated debate on 17 November 1937 it was announced that a further committee, under Lord Cadman, head of the Anglo-Iranian Oil Company, and including Sir Frederick Marquis of Martin's Bank (the subsequent Lord Woolton), would be set up to deal with complaints of inefficiency and 'retarded development' in British aviation.

The Committee reported in March 1938 and did much to substantiate what the critics had been saying. They expressed 'extreme disquiet' at the state of civil aviation in the country. Much of their Report was directed at the management of Imperial Airways, but the network of aerodromes, and the patent unreality of some of the municipal operations which had dutifully been commenced in response to the combined exhortations of Prince Edward and the Secretary of State for Air, also caused the Committee considerable anguish. The aerodrome position should be reviewed, they declared, and a 'co-ordinated scheme' prepared. It sounded, even allowing for the very different circumstances of the late thirties, very much like the appeals for a national airports plan which were to be made by a variety of interests throughout the fifties and sixties; and like them, it failed to awaken an answering enthusiasm in Whitehall. There was nothing wrong with a co-ordinated scheme, the Government said in its reply to the Committee, just so long as it did not go further than advice to local authorities; and so long also, of course, as it did not entail frequent assistance from the Exchequer.

Throughout the war it was clearly seen that there would be a lot of catching up to do once peace returned. It had been appreciated as early as 1936 that to extend Croydon to the size which effective operation would demand was going to be more costly, both in financial and social terms, than could be justified. By 1943 two other contenders for the role of London's principal post-war airport, Heston and Hendon, had been similarly ruled out.

To the evident irritation of some MPs, who were stirred to interrupt with reminders that there was a war on, a small group of aviation enthusiasts made it their business to harry the Air Ministry at every possible parliamentary opportunity. At first, they opposed the cutting down of civil aviation because of the war; after all, said one of them, just because a nation is at war it does not try to shut down the mercantile marine. Subsequently the attack switched to warning the Air Ministry of the need to plan now for post-war civil aviation, while at the same time proclaiming its total unfitness to do anything of the kind.

None was more virulent than Mrs Mavis Tate (Con., Frome), who declared that the only policy of the Air Ministry towards civil aviation had been one of consistent neglect. They were only interested, it was said, in military flying; civil aviation was a Cinderella, to be left in subjection. The war would end, and the Americans (who were now operating flights to North Africa which had previously been a British prerogative) would be found to have everything, leaving Britain with nothing. And where were the airports ready to take the new planes of the future? Not one was equipped for the role it would have to play when peace arrived.

The Air Ministry, it was sadly noted, was not keeping abreast of technological progress. Some members warned that the new jet engines might change the nature of post-war civil aviation very substantially. Others were more attracted by other new technological miracles which they believed to be on the way. There had been speculation before the war about a possible elevated aerodrome on top of the Southern Railway terminal at Charing Cross. This talk sent inventors to the drawing board up and down the land and many radical proposals were produced, some involving the launching of aeroplanes by rocket. The most persistent of these was probably Mr Charles Frobisher's Rotary Elevated City Airport, which was to be created by building a rotating elevated runway supported on top of a high tower, which would automatically align itself to the level of the prevailing wind. The course of the war did not allow some members to be dissuaded from pursuing this idea. On 10 April 1940, on which day the Prime Minister came to the House to make a statement on the military and naval developments following the German invasion of Norway and Denmark the previous day, Sir Reginald Clarry (Con., Newport) was asking the Secretary of State for Air whether he would inquire into this scheme, to which Sir Samuel Hoare replied with regret that his information suggested it was not practicable. On 28 January 1942, as military setbacks in the Far East made inevitable the evacuation of Malaya, Mr Kirby (Lab., Liverpool-Everton) asked the Parliamentary Secretary to the Minister of Works if he would recommend cities such as London, Liverpool, Manchester, Hull, Birmingham, Bristol, Glasgow and Edinburgh to include in their plans for post-war development consideration of the desirability of a rotating elevated airport. Still the advice was that it was not practicable. In June 1943 Mr Kirby tried again; this time the view was that the proposals required further detailed examination.

In fact there was more activity than Ministers were able, on security grounds, to reveal. A committee was set up to make an inventory of aerodromes and consider their future potential. A pamphlet setting out Ministry thinking on post-war runway sizes (municipalities, it said, should think in terms of runways more than a mile long) had also been circulated. 'Thus,' declared the Air Minister, Sir Archibald Sinclair, in March 1944, 'we shall start after the war with an airfield system which was planned, not just grew up, and was planned to take advantage of all developments which the best brains and the greatest experience could foresee.'

What he could not say – although many already knew – was

that the principal airport for London was to be on a site close to the villages of Bedfont, Stanwell and Feltham, where Fairey Aviation had laid out a small grass aerodrome in 1931. Peter Masefield, later Chairman of the British Airports Authority, started work there in August 1935: 'I cycled down the leafy lane from Hayes to Harlington Corner,' he later recalled, 'past the Magpies Inn, to come to the little grass field among the market gardens.' But the little grass field, which had come to be known as the Great West Aerodrome, now had great things in front of it. In 1943 it was decided to develop there a large airfield capable of taking the heavy planes which would be required in the late stages of the war. This airfield could also be used to furnish the first-class peacetime airport for London which everyone knew would be needed. Work on the construction began in May 1944. By August at least one back-bencher had spotted it. 'I have recently seen a most marvellous aerodrome in the course of erection,' said Frank Bowles (Lab., Nuneaton), 'with runways five miles long . . . So there is a policy being worked out administratively and the House of Commons is being kept in the dark.' But Bowles was no civil aviation enthusiast: 'I would like to know,' he demanded, 'whether it is necessary to have another big airport when we have airports already in this country.'

Not everyone agreed that this new airport was the right choice for the principal airport of Britain. But the criticisms of the new Middlesex Airport, and the arguments about alternative locations, tended to reflect local loyalty rather than cool technical assessment. Dr Russell Thomas (Nat. Lib., Southampton) thought the obvious place for it was in Southampton, while Lady Apsley (Con., Bristol-Central) could think of no better site than Bristol. A naval man with a continuing interest in civil aviation, Rear-Admiral Sir Murray Sueter, wanted it put at Portsmouth (it could be linked with London by amphibian planes flying off Chelsea Reach) and added that there should be a second major airport somewhere in the Midlands – 'in the Manchester-Liverpool area', for example. Sir Thomas Moore (Con., Ayr), while conceding that your first terminal must serve your capital city, was anxious to see that the sophisticated wartime aerodrome at Prestwick in Ayrshire should retain its international status. But that too was contentious; Prestwick, declared the Unionist Member for Londonderry, Sir Ronald Ross, would soon be outdated, since transatlantic flights would want to land on water, thus ensuring that the second major airport of the nation would have to be sited on the Londonderry lakes.[1]

The London Airport Planning Committee, which included among its numbers a young, restive and temporary civil servant called Edward Heath,[2] could well afford to conclude that advice of this calibre need not deter them from the work on which they had embarked. They were, naturally, out on an uncharted course. There was a total absence of dependable statistics. The layout panel which designed the ground plan of the airport simply had to make the best deductions it could from wartime knowledge. But in general there was satisfaction with the choice. There were, of course, teething troubles. The work took longer than expected and when the war ended the role of London's first airport had to be assumed for a while by the aerodrome at Hurn, near Bournemouth. And then, what with all the secrecy and novelty of the scheme, some people who were vitally affected did not seem to have been told. The Ministry of Civil Aviation, it was revealed, had failed to tell government departments what it was up to, with the result that twenty houses which had started going up in the site area were now coming down again. (Work on twelve more, mercifully, had got no further than the foundations.)

But such embarrassments were merely temporary. Against them could be set the convenience of a site so near the centre of London. True, there was no rail link, but as the Air Minister, Lord Winster said at the time the airport opened in 1946, 'both the Minister of Transport and I are alive to the necessity of developing access to the airport . . . access will probably be provided by extending the Underground Railway beyond Hounslow West to the Airport itself' – a pledge well on the way to being redeemed when the twenty-fifth anniversary of the opening was celebrated in 1971.

The first proving flight left for Buenos Aires on 1 January 1946 with Air Vice-Marshal Donald Bennett, war hero and former Liberal MP, at the controls. One more essential preliminary was needed – a decision to change the name of the airport from Heathrow to London Airport; foreigners, it had been reported to Whitehall, were exhibiting considerable difficulties in pronouncing the words Heath Row. On 25 March the new airport was officially opened. Lord Swinton, who had become the first Minister of Civil Aviation in October 1944, after pressure from Mrs Tate and her allies had helped finally to prise civil responsibilities from the Air Ministry's grasp, declared at the ceremony: 'I say without the least hesitation that it was the only possible site on which a great airport for London could be built.' While Lord Winster, his successor, added this commendation: 'One of the reasons which led to the siting of the London airport in this district was that it could be done with the minimum of disturbance to householders.'

All vanity of course; in the year of the twenty-fifth anniversary celebrations, complaints from householders about the activities of London Airport (Heathrow) reached a total of 2,359 – and that includes only those made to the airport; others prefer to ring the government departments, the local authorities, the newspapers, the BBC. Certainly the war, commanding an effort in men, money and materials which civil aviation could never have matched unaided, transformed the whole technology of aviation, creating a new breed of aeroplanes which would call for a new generation of aerodromes to take them. Yet the first planes to fly out of Heathrow were not so radically changed from the days when Gandar Dower had hired his aeronautical factotum and when 16,000 people had made the pilgrimage to Croydon to sample the excitements of the new world of aviation. A ten-seater Lancaster was no adequate herald for the screeching jets which tumbled all the earlier assumptions as they moved in on London Airport at the end of the fifties; or for the jumbo jets which, disturbing all the understood ratios between the volume of passengers and the number of air traffic movements needed to accommodate them, wrong-footed the forecasters all around the world a few years later.

It was still by no means clear as Lord Swinton and Lord Winster spoke that a movement towards big planes would be more than a passing phase. There were those who shared the view of Arthur Woodburn, shortly to join the Attlee Government as Parliamentary Secretary to the Ministry of Supply, in a Commons debate in January 1945, that there was no reason for the trend to continue. Were we to believe that one day machines like the Queen Mary would fly across the Atlantic? 'So far as I can learn from the technical people I know the thing is ridiculous.' They would be more the size of a bus; or perhaps of a taxi.

'The aeroplane,' said Air Commodore Helmore (Con., Watford), 'will become the bus of the future.' He, however, was not in this case thinking, like Arthur Woodburn, in terms of size. He was thinking of its general utility. He believed that air travel would become part of the everyday lives of the people. 'It is very easy,' he said, with impeccable confidence if somewhat fallible geography 'to be pessimistic about the future of aviation – just as easy as it was in the old days to be pessimistic about the original railway between Stockport and Darlington.'

This was the kind of contention which politicians were happiest debating. Was flying for the general good – or was it the privilege of the few? That had been one of the classic political divisions over the motor car[3]; and it was as eagerly debated in aviation. Far too much was being made of the prospects of civil aviation, said Frederick Montague, Under-Secretary for Air in the Government of 1929 and Parliamentary Secretary in Aircraft Production early in the war, but now a sardonic commentator on the activities of the coalition, in October 1943. The advantages were for the well to do, not for ordinary people. 'I represent' – he was Member for Islington West – 'some 33,000 of the people of the world, and not one in one thousand of them have ever seen the inside of a plane or is ever likely to.' It was ridiculous, said Frank Bowles, to talk of a post-war boom in aviation; there would never be the fuel to sustain it. Aneurin Bevan, a frequent intervener in aviation debates, noted that people were saying that planes would blossom after the war like the railways. All very well; but they had said just the same about the balloon.

This was the central ground of civil aviation debates; this and the argument between private and public control. Set private enterprise free to build a great competitive business, cried the Conservative back-benches. Nonsense, said Bevan and his allies; they'll be back demanding subsidies – if not open subsidies, hidden ones, like contracts for mail; they will make the profits, and we will carry the industry on our backs. Members of Parliament were not at home in forecasting the future patterns of aviation. Why should they have been? No one was. But with the ancient issues of ideological disputation, they were on much more familiar and comfortable ground.

It was the coming of the jets which, most of all, was to break in on the political arguments with a new, deeply political, yet uncomfortably not *party* political, logic of its own. Their arrival was long delayed. Whittle's breakthrough had occurred in 1936; yet it was not until late in 1958 that they began on regular service from Heathrow.

They were preceded by restrictions. An announcement from the Minister of Transport and Civil Aviation (for Civil Aviation, having cut free from the Air Ministry, was married to Transport less than ten years later) said jet aircraft operating in Britain would be subject to his prior permission, which would not be granted until every attempt had been made by operators and manufacturers to prevent their aircraft from causing 'intolerable disturbance'.

But this was a meagre defence; and it achieved almost nothing. The noise *was* intolerable (an official committee five years later was to declare it so); but the aircraft continued to fly. Nor did those who suffered have any legal redress. The Air Navigation Act of 1920 had prohibited actions for nuisance caused by civil aircraft in flight, and this protection was extended to aircraft on the ground by the Air Navigation Act of 1947. Sections 40 and 41 of the Civil Aviation Act of 1949 consolidated these measures. The reason was obvious. Without this restriction, it was feared, civil aviation in this country might have been brought to a halt.

In wartime, noise had not been an issue. Indeed, it had, if anything, been reassuring. Not only was it inevitable in wartime; when it was *our* noise, not the enemy's, it was a heartening sign that our military effort was being maintained. In the immediate post-war years, that attitude diminished – though slowly – and in any case, the noise then was, by modern standards, mostly mild. There were scattered complaints in Parliament, notably from MPs like Sir Alan Herbert, who lived in Chiswick, and Sir George Harvie-Watt, who represented Richmond. It was – and is – those areas, a little distant from the airport, which generate most complaint. Many living around Heathrow today have gone there voluntarily, perhaps because of the airport and the employment it has created; or perhaps despite it, because other considerations outweigh the burden of noise. It is the peripheral, and usually better-off, areas which are deaf to the advantages of the jet, and are, they claim, being progressively deafened by the unavoidable disadvantages.

In 1960, recognising the growing public objections to uncontrolled noise – not aviation noise alone, though that was a very important part of it – the Conservative Government set up a committee on noise under the chairmanship of Sir Alan Wilson, an industrialist with a scientific background who was a favourite Whitehall choice to chair committees. Their Report, published in March 1963, declared:

> We are agreed that the noise to which many people near London (Heathrow) Airport are subjected is more than they can reasonably be expected to tolerate. This situation has arisen because no one, either in Britain or elsewhere in the world, was fully apprised of the great increase in annoyance which would be a consequence of the introduction of large jet aircraft. As a result, Heathrow has proved to have been

3

established in a much too densely populated area, and no good solution to the noise problem is possible.

So much for 'the only possible site' on which a London airport could be built. So much for the 'minimum of disturbance', now seen to be an intolerable affliction. But what could be done to alleviate it, since it surely could not be removed? In the short term, said the Wilson Committee, there could be palliatives, such as grants to help pay for soundproofing. In the longer term there could be more foresight. Early and firm decisions must be made on future airports, and once these decisions were taken, there must be strict controls to prevent development in areas which might be intolerably affected by noise. It was also high time said Wilson, for the Government to start thinking about the noise problems which would be created by supersonic flight.

The Government rejected the palliative proposal; they were not satisfied, said Lord Hailsham, who was Minister responsible for Science, that the situation at Heathrow justified what Wilson was saying, and in any case there were overwhelming difficulties in deciding how to draw a line between homes eligible and in-eligible for soundproofing grants without creating inequity. When Roy Jenkins reached the Ministry of Aviation in 1965, this ruling was reversed. The administratively impossible was performed – though the predicted difficulties duly occurred. In the same year, a qualified night curfew was imposed, restricting the number of movements by 'pure jet' planes within a stipu-lated slice of a summer night.

That was one lesson learned in the twenty years between Lord Winster and Roy Jenkins. But there was another effect of the creation of the modern airport of equal significance to the peace and equanimity of the lives of people living around it, which took even longer to be digested.

There are two serious consequences of airport development which have only slowly been understood. They both result from the ripple effects of siting a landing strip in the middle of a stretch of countryside. The first is the physical effect – not just on the area where the airport is built, but on a huge swathe all around it. The second is the economic effect: the creation, not just of thousands of new jobs for people dependent on the airport, but of a host of other new industrial openings created by those industries which home on an airport like a moth on a flame. Even at the time of the Stansted debate, these were perceived only dimly – and by Whitehall, more dimly than by some others. As will be seen, the Government were to call not one witness on regional planning issues before the Stansted inquiry; it was the opposition which brought in experts to testify what an airport can do to feed an economically depressed area or to overheat a prosperous one.

The ripple effects of an airport were described many times before the Roskill Commission, sometimes with that element of exaggeration which comes from using evidence to plead a pre-determined case. But they were graphically set out in the evidence of the consultant to the Wing Airport Resistance Association, Professor Tom Hancock. The mark of the airport, he said, will be left on three concentric areas. First, there is the airport site itself; initially, in the case of Cublington, this would be five miles by three miles and everything in it would be destroyed. Three villages would disappear completely; some 1,400 people would be displaced; twenty-nine buildings listed for architectural or historic merit would be lost.

Second, there would be the area, a further rectangle around the area of destruction, the nature of which would be changed by noise. This second area could in turn be subdivided into those parts where the noise would be intolerable and those where it would merely be oppressive. In the 'intolerable' zone – technically, the zone with a noise rating of more than 50 NNI[4] – lived a further 8,000 people, with a further 219 listed buildings scattered among them. This area, said Hancock, would become a 'no-man's land' between the airport site and those places outside which, though severely harmed by the airport, would still be capable of 'normal activity'. Add in the 'oppressive zone' – that with noise ratings between 35 and 50 NNI – and you had an area forty-five miles by six containing something like 70,000 people, all of whose lives would be sharply if not utterly altered by the coming of the airport.

But even beyond that there would be an area which Hancock called 'the area of consequent urbanisation'. The noise in this wider zone might not be overpoweringly objectionable; yet the injection of a modern airport, requiring a supporting population of perhaps half a million people, into a piece of countryside, would gradually alter the character of this zone too.

If you drive around the lanes along which the villages of north Buckinghamshire – or the Stansted sector of Essex – are threaded, it is not difficult to see what this entails. The airport will generate traffic, first in its building, then in its operation. It will attract certain kinds of manufacturing industry; it will also create a market for hotels. They will draw in service industries, like garages, long-stay car parks, and manufacturers of vehicle parts.

If you drive along, say, the beautiful road which leads from Stansted through Elsenham to Thaxted; or along the delightful lane which bounds across the little hills from Weedon to Aston Abbotts, south of Cublington, you can imagine what this means. Soon anyway, perhaps, the natural growth of the motor car is going to lead to claims that these roads are no longer adequate for the traffic they carry, and ought to be straightened and widened. But the enormous multiplication of local traffic which a big planning development such as a third London airport will bring is likely to bring that day much closer. What might at present be avoided would come close to inevitability. Even if noise did not ruin the rural tranquillity of such places, the urbanisation, the uncontrollable generation of more and more traffic, could be guaranteed to do so.

Or drive west of London Airport, beyond the boundary road – now converted into a fast link for traffic from south-west London and the M4 – and sample the little area of patchy private housing and market gardens which has still not quite been eaten away by urban spread. This must have been what this sector of

Middlesex was like before Heathrow. It is not beautiful; probably it never was. But it is some kind of touchstone of what the airport has largely obliterated. And then, drive south-east, into the suburban agglomeration of the airport's dependencies . . . Transfer this kind of contrast to the more beautiful natural circumstances of Stansted, of Cublington, and indeed of the eccentric, endearing country lanes north and north-east of Southend, and you will see what airport urbanisation there could mean. Urbanisation, unlike noise, has yet to meet its Wilson Committee or its Roy Jenkins; but it is an inescapable factor in any decision on where a modern airport – unlike the *ad hoc* airports of Captain Gandar Dower – should be sited.

3

. . . In which Gatwick is dispensed with, and then immediately reclaimed; though for what purpose none can clearly say

Heathrow – the name tended to persist, despite the difficulties of the French in wrapping their tongues around it – was one of seven London airports. The others were nearby Northolt, Blackbushe (on the A30 near Camberley), Croydon, Gatwick in Surrey, Stansted in Essex and Bovingdon, near Hemel Hempstead, in Hertfordshire. Before long, as the skies grew busier with civilian planes, it became clear that this system could not be allowed to continue. It wasted resources; and it created unnecessary problems of air traffic control. A committee was set up to rationalise the London flying pattern. The result was a White Paper published in July 1953.

It began with a stirring Imperial call delivered in language which would have gratified Mrs Tate and those of her colleagues who had so fervently condemned the lack of patriotic devotion to British civil aviation of the Air Ministry during the war. London, it proclaimed, was the most important centre in Europe for travel and the transit of goods by air. Within the next ten years, the number of scheduled flights in the London area might double.

The increase of air traffic [said the White Paper] presents great problems, which must be solved if Britain is to maintain the outstanding place in civil aviation which she has already won in the face of keen international competition. In the years to come, the air will be hardly less essential to our well being than the sea. To no small extent, the future of this country in world trade and as a great power will depend on our holding and indeed advancing our place in this form of transport.

In good weather, London Airport ought to be able to deal 'for many years' (the forecasts were no more precise than that) with all expected traffic bar some charter flights. But not in high summer; an additional airport would be needed then to take some of the traffic now operating at the height of the season from Northolt, which the Government intended to close for civil aviation and restore to the RAF.

There was also the problem of bad weather diversions. The answer was to choose an alternative to London Airport, and a supplementary airport to stand in for that when necessary. For unless the problems of diversions and of seasonal traffic could be solved, international traffic might turn away from these islands. 'And this,' said the White Paper, 'must at all costs be prevented.'

Fifty candidates were examined, and from them, a short list of four was compiled. Three – Gatwick, Stansted and Blackbushe – were part of the existing network of seven; the other was Dunsfold, in distant Surrey, almost forty miles from the centre of London against Stansted's thirty and Gatwick's twenty-seven.

The choice fell on Gatwick, which was found to be, if not the 'only possible' site, at least the only suitable one. It was, on the face of it, a surprising choice, not least for the people who lived around Gatwick, in the little villages of Charlwood and Lowfield Heath, since it was only four years before that the same Ministry of Civil Aviation which was now carving out such a glamorous role for Gatwick had been ready to consign it to the civil aviation scrapheap.

The sequence of events was this. The new Labour Government of 1945 had early in its life announced its decision to take both the principal airlines and the principal airfields into state ownership. The first of these proposals had been deeply controversial. To some Conservatives, indeed, it was little short of a national disaster. The former Minister of Transport, Leslie

Hore-Belisha, for instance, warned that we might go the way of Spain, whose decline as a great power, he declared, quoting the naval historian Mahan, was due to the way in which its government ran the whole of the nation's shipping itself and 'cramped and blighted free enterprise'.

The second, however, raised hardly a murmur. The reason was plain. 'The nationalisation of the airfields,' Ivor Thomas, Parliamentary Secretary to the Ministry of Civil Aviation, noted later, 'has met with very little opposition because the cost of these airfields is so great that no private organisation is anxious to have the responsibility of running them.' The White Paper of December 1945 (which cost twopence) itemised the argument for state airfields: big spending was needed to match them to the great modern aircraft they would need to take; many airfields were already in state hands through wartime takeovers, and for the state to hang on to them would be an economic solution and a simpler one than derequisitioning. In any case, the state had for many years looked after meteorological, radio and air control services: in the changed circumstances of post-war Britain it was a natural development of that policy to put the airfields into state hands too.

There was to be a national plan. 'If we are to secure the orderly development of transport aerodromes in the right place and up to the right standards,' Ivor Thomas told the House of Commons in January 1946, 'it is necessary to have a central plan.' Yet a national plan was, in a sense, no more than a stopgap. It was the committed policy of the Labour Party that post-war aviation should be international. But since – on the evidence of the great conference on post-war aviation staged at Chicago in 1944 – the nations were not yet ready for this solution, Britain would work out a national plan which, in time, would slot easily into the superior framework of an international policy on aviation.

This too had been a staple argument of the war years. Frank Bowles, as a Labour aviation spokesman in the coalition days, advocated the creation of World Airways Limited. In subsequent debates he developed this idea; the Airways Corporation would be absorbed in time into a World Communications Authority which would also comprehend shipping and rail. Bowles further infuriated the Conservative benches by a suggestion that the running of this authority should be left to the smaller nations, such as Switzerland and Sweden.

This proposal reflected in part the internationalism of Labour; but in part it was an attempt to draw a lesson from the build-up of military power before the Second World War. Labour MPs sometimes quoted Stanley Baldwin (not necessarily a persuasive source for those days): 'civil aviation', he had said in 1932, 'is a thing to be feared – every plane engaged in it is a potential bomber.' 'The Lufthansa,' declared Edgar Granville (Ind., Eye), 'was the father of the Luftwaffe.' And Aneurin Bevan described civil aviation as 'international dynamite'. The Conservatives scoffed. 'The international bird,' said Hore-Belisha, 'will not fly.' But in the wake of the allied struggle in the war, and of the landslide victory of Labour in the 1945 election, it did not seem so impossible then. A Commonwealth conference called by the Government had declared unanimously for an international air transport authority.

Meanwhile, there was at least some measure of international co-operation resulting from Chicago. Each country was to designate one or more airports as 'international'. Britain's would be two: London and Prestwick. For the rest, there would be a network of regional services, built on a pattern of airfields selected as of greatest potential.

There was plenty to choose from. Britain had been described, both contemptuously and fondly, as an 'aircraft carrier' in wartime, and official statistics showed why. The Air Ministry, said Sir Archibald Sinclair in February 1944, had been carrying out the most gigantic civil engineering and building programme in the nation's history; it had laid down concrete track 'equivalent to a thirty-foot road running from here to Peking'. There were 640 airfields in use during the war, covering something like a quarter of a million acres. (For comparison, the area of Greater London, in 1973, was 394,000 acres.) An Airfield Selection Committee was set to work assessing the prospects for the better airfields among them and in 1947 produced a list of forty-four. The selection did not cover the London area, which was served by its network of seven. But there was already rethinking about these airfields too.

Gatwick, for one, did not seem destined to last much longer as an airfield of any importance. In 1946, when it was taken from the military, it was decreed that it would only have a very restricted use. In view of its earlier history, that was, perhaps, not very surprising. First licensed in 1930, under the name Cotland Farm, it had expanded in 1934 and in 1936, it was named, along with Gravesend, as a bad weather substitute for Croydon. But in 1937 British Airways had abandoned it for Heston because they found the Gatwick weather too unreliable. The 1946 decision came as a great encouragement to the Ministry of Town and Country Planning, which had for some time been eyeing nearby Crawley as a possible site for a new town. Now, with this assurance that Gatwick would not expand, they were able to give the go-ahead.

No sooner had the Order authorising the development of Crawley been made in February 1947, however, than the Ministry of Civil Aviation changed its mind. Later that year it was found that plans were being made to extend the airport whose use it had restricted only a year before. The new town Development Corporation moved in quickly to register its total opposition, and by 1 March 1949 the threat seemed finally to have receded.

Gatwick and Blackbushe [George Lindgren, Parliamentary Secretary to the Minister of Civil Aviation told the Commons that day] are both requisitioned aerodromes and it is not the intention of my Ministry to acquire either of them. Gatwick is being derequisitioned in September this year and we are hoping to derequisition Blackbushe by the end of 1950. To make facilities for the charter operators who will be displaced from Gatwick and Blackbushe, we intend to develop Stansted Airport in Essex which is near Bishop's Stortford, and in developing that airport, we hope to make it available for diversions from London, which may give an opportunity, perhaps, at a later date, for us to give up Bovingdon as well.

So the future was clear. Stansted was the coming airport; Gatwick, Blackbushe and probably Bovingdon had been chosen for the discard pile. But the Ministry was not, it soon appeared, certain about that either. In 1950, Crawley Development Corporation sought the aid of the Ministry of Town and Country Planning to fight off a £2·5 million development plan for Gatwick. And from then on, the threat to Crawley continued to grow, until in July 1952, the Ministry of Town and Country Planning retired defeated and the Ministry of Civil Aviation announced that Gatwick was indeed to be developed.

The 1953 White Paper, then, was confirming a choice which had been made well before its arguments were published. There would be scope for a two-runway airport, with possibly a third later on. Six hundred acres would have to be taken and sixty houses demolished in the first stage of development. Certainly there would be disadvantages; noise, for one. 'It cannot be denied,' the White Paper conceded, 'that aircraft are noisy and that in this densely populated island the construction of aerodromes inevitably means some disturbance.' But the airport would be planned to reduce noise and disturbance generally as well as to minimise the risk of accidents. Then there was the cost: £6 million, it was estimated, over seven years. (A year later it had already risen to £10 million.) But there would be ways of off-setting that, including the sale of Croydon.

But what, people asked, and especially those people for whom Gatwick inevitably meant some disturbance, had happened to the airport near Bishop's Stortford which Lindgren had marked out for greatness in his statement four years before? Well, Stansted, it appeared, was on the wrong side of London (a factor which had not come into the calculation in 1949) and, with a travelling time from London of up to two hours, it was considered inaccessible. But most of all, it could not be proceeded with because of considerations of national defence: 'there is at present a very large amount of military flying from nearby airfields which severely limits the use which can be made of it by civil aircraft.'

Counsel for the Ministry at the Gatwick inquiry was very sorry about this abrupt change of plan but it appeared that the people to blame were not the Ministry but the Russians. When the earlier assurances had been given, he said, it was believed that a period of peace was likely to be ahead. Since then the re-armament programme had commenced and Stansted, the hoped-for alternative, had become 'practically unusable' for civil aircraft.

That seemed conclusive. Stansted was now out of the running. Instead it was Surrey and Sussex which were in the firing line. There was, of course, the local inquiry; but it was one which put opponents of the scheme in various kinds of difficulty. First, there were the terms of reference, which seemed to have been specifically designed to stop people talking about some of the things which concerned them most. The Inspector, Sir Colin Campbell, was charged with the task of 'hearing and reporting on objections relating to the suitability of the site or to the effects of the proposed development on local interests'. In addition, it would be open to objectors to suggest detailed modification of the scheme now proposed. 'It is not however open to them to question the need to provide a major airport to serve as an alternate to London Airport nor to make proposals as to how that need should be met.'

In other words, if you wanted – as did the county councils, eager to be constructive – to press the superior claims of somewhere far away, you would be out of order. If you wanted – as at least one witness to the inquiry had come prepared to do – to argue that the airport would be a white elephant because of the replacement of conventional traffic by the helicopter, that would not do either.

But there was also grievous difficulty in understanding exactly what the purpose of the new Gatwick Airport would be. Was it to be a major civil airport, an established number two to

Heathrow's number one? Or would it be an establishment of a more occasional kind?

The White Paper appeared to carry a specific promise: 'Gatwick will not be used intensively all the year round. Its principal purpose will be to receive aircraft diverted from London Airport when visibility is poor there and this purpose could not be achieved if too many regular services were based at Gatwick.' That would have made the prospect of an airport more acceptable, if only people had been able to believe it. But there had been confusion, or ambiguity, about it. The local MP, Frederick Gough, had taken the obvious course of asking the Parliamentary Secretary at the Ministry of Aviation, Reginald Maudling, to make the matter plain. But the exchange had only solidified the mystery.

> We consider [said Maudling] that Gatwick will serve two useful purposes: one as an alternative when aircraft need to be diverted from London Airport in bad weather conditions, and also for the secondary purpose of providing a base from which could be transferred a certain amount of the regular load now carried at London Airport.

Which was clear enough, if only one knew what exactly was meant by 'a certain amount'; but that remained as much of an enigma as before.

Suspicion, at any rate, persisted. Nor was that all. The county councils, the residents' associations and the private individuals who lined up to do battle with the Ministry at the public inquiry discovered that one crucial member of their forces had gone absent. Crawley Development Corporation, which had so often implored the Ministry of Town and Country Planning to prevent this pernicious proposal, was now to be found quietly registering its willingness to accept it. Reluctantly, of course – the Corporation, said its chief executive, Colonel Turner, still considered that the juxtaposition of the airport and the new town was bad planning. But they now believed that the national needs of civil aviation must override such considerations.

Very public-spirited, in the widest sense; and an attitude certain to commend itself to the Ministry of Aviation, whose sense of priorities so closely accorded with it. But the local authorities suspected dirty work at the cross roads; a suspicion strengthened when they discovered that various crucial pieces of information about the expansion scheme, denied to them, had been whispered readily into the complaisant ears of Crawley.

So they did what they could to highlight the unwelcome nature of the noise (the equivalent, said one local man who obviously did not realise what the jet engine had in store for him, of a busy road running through your village) and the weather, which could not have changed so very radically, it was asserted, since it had disposed of British Airways in 1937. The Inspector found some of the objections impressive; noise effects on Charlwood and Lowfield Heath, he said in his Report to the Minister 'merit your consideration as a substantial point against the proposals'. He was unhappy with the evidence of Captain Vernon Hunt, the Ministry's expert on air control; this evidence should be carefully re-examined for the rigidity with which some sectors around London had been excluded.

And he was clearly distressed by the plight of some of the objectors; people like the proprietor of Sam's Halfway House (who was seventy-two) and the man who had recently paid £8,000 for the Tea Cup Café, both expecting to make a comfortable retirement out of the traffic on a Brighton road which under these proposals would be routed somewhere else. He made a special plea for a new look at the compensation laws; a plea in which he was to be echoed on different, but comparable, grounds by Roskill nearly twenty years later.

But that was as much as he felt he could do – certainly within the tight terms of reference which the Ministry had drawn for him. It is the privilege of governments to define the rules under which their decisions are challenged at such inquiries, though they do not often weight the decision so heavily in their favour as they did in the Gatwick case.

And so, in a White Paper in October 1954, the go-ahead was given for the development of the new Gatwick Airport. But the airport it envisaged sounded rather different from the mild conception of occasional use which had been floated before the previous year's inquiry. Gatwick was baldly described as 'a second main civil airport for London'; exactly what the objectors had suspected all along that the Ministry intended it to be.

The awkward, disruptive change of plan, the unfavourable public inquiry brief, the overriding influence of defence considerations, the picture created before an inquiry, only to be removed and replaced by something less enticing when the go-ahead was given – the story of London's second airport was indeed a sadly mismanaged affair. That, as the airport began to take shape, was widely and generously admitted. It was, however, a foretaste of what was to follow.

4

. . . In which a National Plan is devised, and promptly disappears, to the great inconvenience of those who come later

After Gatwick, no major decisions were taken on London's airports for more than ten years. Yet these were formative years in helping to create the conditions in which the subsequent decisions on Stansted, Cublington and Foulness had to be taken, and they help to explain some of the frustrations which the search for sites entailed at the time of the Roskill Commission.

Above all, it was at this time that the national airports policy which the Labour Government had set out to construct in the immediate post-war years began to shrivel away and finally, unmourned and unremarked, gave up the ghost. It was done without any conscious decision that this should be so. Even the Conservative Government which came into office in 1951 did not make the burial of this vision of state enterprise the occasion to celebrate our escape from the fate which, in Hore-Belisha's view of history, had overtaken Spain.

The retreat from planning began because there was not enough money to go round. The Labour Government of 1945, like its latter-day successor, found itself forced to whittle down and even abandon cherished commitments because of straitened financial circumstances. The economic crisis which was building from the time Labour took office and which precipitated the devaluation of 1949 was also the mortal enemy of a rational, premeditated national airports system. Services between regional centres – a key element in the original plan – lost money and were shut down. Only five months after the 1947 plan was announced, George Lindgren was having to tell protesting MPs that services

4

between Bristol, Cardiff and Southampton had had to be withdrawn as it was not possible to carry the financial losses they entailed. Prestwick-London had been cut out too 'because of the requirement to retrench'. A month later he was back at the despatch box explaining that because of cuts in the capital expenditure programme, schemes for airfield construction were having to be delayed.

Civil aviation continued to lose substantial sums of money, and the Treasury was harrying the Ministry to staunch the flow. Any hope of operating the kind of network of services envisaged in the original plan was dead within a year of its publication.[1] Ministers' answers to the anguished cries of provincial members, whose constituents now found themselves off the airways map with little prospect of getting back on, took a noticeably sharper and more irritated tone.

It was the provinces which were going to suffer. London, with its more developed market, would be looked after as usual. To provincial members that was inequitable. To Ministers, conscious of the need to save every possible penny, it was simple economic common sense. When Mr Wilkes, the Member for Newcastle Central, drew attention to the plight of Tyneside, which had been promised an airport at Boldon in the plan but saw little chance now of getting it, the Treasury bench were ready for him. 'It is a matter of priority', Lindgren told him. 'Which shall be first, Jarrow tunnel and houses, or the development of the airfield?' This answer seems to have been considered so telling that when another North-East Member raised the same question in March 1949, he was immediately hit over the head with the Jarrow tunnel too.

North-Eastern MPs, and their counterparts in other regions, protested vigorously at the loss of long-term benefits for short-term savings in this way. 'We are a development area,' said Fred Willey (Lab., Sunderland), 'endeavouring to attract into the area new industries which are likely to make extensive use of air freight. We have no shortage of manpower and materials.' So why, having been denied a projected airport at Croft, were they now being refused the promised development at Boldon? Ministers, no doubt, saw the force of the contention; but in terms of what they were able to spend one might just as well ask for an airport on the moon.

The long-term effects of this retreat are hard to calculate. But when, in the early sixties, regional planning suddenly became a vogue, and Lord Hailsham, in a cloth cap, stalked the workless streets of Wearside and Tyneside at the behest of Prime Minister Macmillan, the lack of a good air-base was widely recognised as one of the handicaps from which the region had suffered in attracting industrial investment. It was the usual vicious circle. Because there were not enough industrialists to make a strong enough call for air services in the North-East in the late forties, the provision of a North-East airport sank rapidly to a low place in the list of priorities. Subsequently, because there was no big airport, industrialists who might have been attracted stayed away, congregating in those areas where air services were readily accessible. The result of this was that London and the South-East where air services were viable, became more of a draw than ever; so that by the time the third London airport was being discussed, London could accurately be described as the 'honeypot' of air demand. A lot of that was due to non-business flying, of course; tourists have always been more eager to inspect Buckingham Palace and Windsor Castle than the glorious coasts and rivers of Northumberland or the bleak grandeur of Hadrian's Wall.

Looking back, it is easy to suggest that a little more foresight in the post-war years would have relieved some of the heavy demands which the regions made on the economy later on. But it cannot be proved either way. And in any case, this was a time – there have been a good many in recent political history – when foresight was too expensive a commodity for the government to invest in.

And so, the London airports' map having been rewritten, the provincial airports' map was gradually rewritten too. The Ministry began to look around for people – whether local authorities or private interests – who might be willing to take local

aviation off its hands. Manchester was an early arrival in the
queue, and in 1950 Ringway was taken over by the city. A procession of airports – Shoreham, Tollerton, Lympne, Edinburgh
Turnhouse, Weston-super-Mare, Yeadon – were surrendered in
the next five years.

After the 1948 Berlin airlift, the original commitment to state
ownership and state operation was significantly relaxed. Private
airlines were increasingly permitted to operate scheduled services.
And new airports from which to operate these services were
developed – some, like Southend and Newcastle, by local
authorities and others, like Exeter, by private operators.

'By the middle fifties,' wrote Rigas Doganis, in his 1967
Fabian Tract, *A National Airport Plan*, 'hopes of enforcing any
kind of national airport plan were gone. Not only had the
Ministry divested itself of many airports it had originally owned,
but it also failed to use its limited planning powers to any purpose.'
The Civil Aviation Act of 1949 gave statutory powers to the
government to control airport development, but these were
sketchy and permissive: 'it became easy and fashionable to
establish a local airport and many authorities did just this.' The
result was the direct opposite of what had been intended in 1947.
Some large areas had no airport at all, while others – and this
became the situation in the North-East – had two competing for
the same traffic.

The anomalies of airport geography were matched by the
anomalies of airport proprietorship. The British Airports Authority, created in 1965, took Heathrow, Gatwick, Stansted and
Prestwick. All but Stansted (whose future was still in dispute)
could clearly claim high national status. But some other airports
which made an equally powerful contribution to the total airport
picture – and certainly a much more significant one than Stansted – remained in private or local authority ownership. Take as
one test the total of aircraft movements at each airport. On 1966
figures, Heathrow and Gatwick were certainly the two top contenders in the league. But Prestwick was seventeenth, and Stansted
fourteenth – way behind Edinburgh, Manchester, Birmingham

and even Blackpool, which were runners-up to the two main London airports. Or take the total number of passengers handled; on this test, Manchester, Glasgow, Belfast and even Southend would seem to have a greater claim. In historical terms, the reason for the division of responsibility was clear enough. It would have been hard to say to Manchester: you have built up such a busy airport that we shall have to take it away from you. But in practical terms, and above all in planning terms, the dispositions hardly made sense.

Or take the situation in Scotland. 'It is somewhat odd,' said a witness before the Select Committee on Nationalised Industries in February 1970, 'that three airports in close proximity [Glasgow, Prestwick and Edinburgh Turnhouse] are operated by completely separate organisations, one by the British Airports Authority, one by the Glasgow City Fathers and one by the Board of Trade.'

James Barnes, Under-Secretary, Civil Aviation Division 3 at the Department of Trade and Industry, told the same Committee about another curious anomaly, created by historical circumstances and still unrationalised. At Birmingham, the Government was able to control spending plans for the airport because it paid sixty per cent of the cost of all approved schemes: 'this is a most exceptional arrangement and one which I personally wish did not subsist today.' Birmingham's long West Midlands' civil war with Coventry for the dominant place in West Midlands aviation had long been notorious, with Birmingham gaining in passenger and aircraft traffic and Coventry doggedly refusing to concede defeat.

'Practically all the municipal airports are running at a loss,' Goronwy Roberts, Minister of State, Board of Trade had told the Select Committee on Nationalised Industries earlier. 'Yet the local authorities appear to be ready to bear a loss on the rates.' Or more accurately, it was the ratepayers who were bearing a loss on the rates; indeed, ratepayers in places like Coventry could claim to be picking up some formidable bills for civil aviation traffic to stick behind their clocks, since apart from their subsidy to air travellers at their municipal airport (estimated by Doganis in

1967 at £3·13 per passenger) they also contributed as taxpayers to Board of Trade costs and other government subsidies. For in certain circumstances, a municipal operator could get a government subsidy too. 'We do, rather grudgingly, rather cheeseparingly, I admit, very rarely, make grants to local authority aerodromes,' James Barnes explained to the Select Committee. 'We did make a grant to Newcastle, we did make a grant to help Teesside get established.' Barnes, it should be added, did not accept the common charge that the two North-East airports were an unnecessary duplication. But if they were starting with a clean sheet, he said, he doubted if they would want airports both at Glasgow (Abbotsinch) and Prestwick; at both Birmingham and Coventry; even perhaps at both Manchester and Liverpool. It was perhaps arguable too if they would have wanted airports both in Bournemouth and in Southampton.

Indeed, municipal aviation, to which councils up and down the country had been so effectively exhorted by the Prince of Wales in 1933 and towards which they had been more discreetly pushed by indigent governments of the post-war years, had not turned out to be the everyday feature of Britain's national life which enthusiasts in the war time years had supposed. Then, Wolverhampton's MP had roundly asserted the determination of his local authority not to be dependent on the airfield of its larger rival, Birmingham, which was at least twenty miles away. He had assumed as an unchallengeable fact that a man who a few years later took it into his head to travel from Wolverhampton to Hull would automatically choose to fly. But that has not happened; although cross-country communication by rail has been made an even more perilous affair by the Beeching reforms than it was in the days of Bradshaw, the switch from rail to air travel which was once expected has failed to materialise; and indeed, technological developments on the railways have allowed them to claw traffic back. There are perils in expecting too much from new technologies, as well as too little: 'we have to plan far ahead,' Ivor Thomas had said in 1946, 'but we must learn from the example of Thales who, while looking at the stars, fell into a well.'

In terms of civil aviation planning, successive governments have certainly not been guilty of looking at the stars; they were, however, as will be seen, going to fall into a well – even a series of wells – anyway. But the collapse of the 1947 plan was only part of the difficulty. For civil aviators form only part of the army who compete for the space in the sky. There is also military aviation, as well as test flying, gliding, and other subsidiary activities. And in the fifties, those years of no major decision, there were many minor decisions in this field too which were to loom as formidable as a single major one. Sites were chosen in places, each of which was convenient for the stipulated purpose, but which, had there been some national plan, would undoubtedly have fallen under the planner's veto.

The clearest example of this was to appear at the time of the Stansted controversy. When the Stansted decision was re-examined by an inter-departmental committee between 1966 and 1967, the roving eyes of the Government's investigators lighted on the Silverstone area of Northamptonshire. It was immediately noted that it had a number of advantages. As Anthony Greenwood, Minister of Housing and Local Government, told the Commons later, the proposition had seemed to him on planning ground to be 'most attractive'.[2] Unfortunately, however, it was also most unworkable. New bases for military flying had recently been built in the area, and the choice of Silverstone would have shut eight RAF stations completely and the use of thirteen more would have been severely limited. And Greenwood drew from this a moral which must rank among the most accurate diagnoses of what went wrong with government policy in those years.

It is quite clear, looking back, that at least from 1953 onwards the assumption was consistently made that Stansted would be the third London airport. That assumption helped to determine the routeing of air traffic, including military traffic – honourable members must remember that there are twice as many military movements as civil movements every day – and also helped to determine the distribution of military airfields and other installations.

So that was that. Silverstone was put back on the shelf, where it continued to glitter all the more alluringly for being so heart-rendingly unattainable – especially since there would never be one of those awkward inquiries at which governments are required to justify in cold everyday terms the qualities which they see in the objects of their current passions.

By the time of Stansted, and certainly by the time when Roskill at last looked at airport planning in the necessary depth, hundreds of people, making individually sensible decisions, had contrived to manufacture a kind of aeronautical assault course which every ambitious airport planner must be required to attempt and at which many, not surprisingly, fell.

On this score, at least, Roskill and Buchanan could join hands.

One matter which emerged most clearly [said the majority report] from even a cursory study of the quarter of a century which had elapsed since the Second World War was the absence of any national airports plan in this country. Subject to obtaining the necessary planning consents, which seem to have been granted by reference more to local than to national considerations, any person or body of persons who believed that an airport was an economically viable or simply a desirable proposition in a particular place has been able to convert that belief into action.

Roskill had been encouraged by some witnesses – he gave as an example the Confederation of British Industry, though they were not alone in this hope; it was shared, as will become apparent, even by the Board of Trade[3] – to produce a national airports plan. Certainly not, he replied: 'it was no part of our task as defined by our terms of reference to devise such a plan and we have resolutely refused any attempt to do so.'

Buchanan declared that post-war planning policy in Britain, in which he found much to admire, had contained one serious deficiency. There had been no policy on airport location. In the years after the 1947 Town and Country Planning Act, statutory undertakers and nationalised boards had been left largely outside the scope of land use planning. The situation had been aggravated in that

civil aviation was for some years a more or less parentless child handed around from one ministry to another. The policy, if such it can be called, seems to have been based on the 'Have a Go' principle, that is to say any public authority or indeed any private person has been at liberty to have a go at establishing an airport. It is true that planning permission has been required, but this would be involved only with fairly local issues, and air traffic control licensing procedures would have to be gone through, but the principle has re-mained that if some body or person thinks he can make a go of an airport, he has been at liberty to try. The result is that airports have come and gone. Some, by accident of location more than anything else, have picked up custom and grown, as at Manchester and Luton; some have struggled for a time, and then given up, as at Coventry; others jog along. Practically all those that have survived are badly situated, being tucked into re-entrants in closely developed urban areas, and causing widespread nuisance as a result. Of an overall location policy properly integrated with land use planning, there is precious little evidence.[4]

As they tried to cut their way through the accumulated under-growth of past decisions and unformulated assumptions, Roskill and his team were constantly falling over other airports which were crying out for decisions to be made but which had been kept waiting – partly because of the creation of his own commission.

Gatwick, for one. Was Gatwick due to have a second runway built, or was it not? The British Airports Authority said that it was, and they quoted Douglas Jay speaking in a Commons debate on Stansted, as their authority. Not good enough, said Roskill: 'it appears to use that what was said by the then President of the Board of Trade was in the context of a debate on the Stansted issue and cannot in any way be regarded as a binding commitment on the present government.' So far as the Commission was concerned, the future of Gatwick was one of the great intangibles.

Then there was Luton. Luton was one of the success stories of municipal airport operation. Its growth had accelerated sharply in the late sixties, and was set on a course of further acceleration when Roskill reported. Luton had decided to acquire the airport in 1934 and had officially opened it as a municipal airport in 1938, just in time to see municipal flying halted because of the war. The 1953 White Paper had not seen much hope for it. It created problems of air traffic control and would need considerable and costly development.

This however was just what it was going to get. The Inter-Departmental Committee which chose Stansted in 1961–3 looked on it with no more favour than their predecessors of ten years before. Not suitable for a major airport, they declared, and it would have to cut down its ideas once Stansted got going. But Luton was not deterred by this lack of Whitehall encouragement. A concrete runway had been laid in 1960 and from then Luton's operations never looked back. Some £2 million had been spent by 1965; by 1969 that figure had been doubled and passengers, mostly bound for continental holidays, had passed the one million mark.

There were agonised public protests – there were a good many at Roskill's hearings – about the noise which was involved. But in commercial terms the effects were extremely rewarding. Passenger traffic was up 67·3 per cent in 1968 as against 1967; in the following year, it actually doubled. It was up by thirty-two per cent the year after and by nearly forty in 1971. Luton now stood third in the national leagues, after Heathrow and Gatwick. Aircraft movements at the airport, which passed 50,000 for the first time in 1969, had reached 70,000 in 1971, when Luton lay fourth in the chart behind Heathrow, Gatwick and Glasgow. It was exactly the kind of record which Whitehall encouragement of municipal enterprise had been designed to produce, and all the more valuable for being so nearly unique. Little wonder that some witnesses before Roskill asserted that, in Luton, London had its third airport already, and that Roskill's search was really designed to uncover airport number four. Little wonder that Luton cropped up again and again in the marshalling of evidence and the submission of witnesses; and that the boost which Luton was expected to get from the siting of an airport at a place as distant as Foulness became a perpetual subject for debate. 'This suggestion attracted so much attention at the final series of public hearings,' commented Roskill wryly, 'that at times we devoted more attention to the future of Luton Airport than to the siting of London Airport.'

Luton was planning to develop its goldmine still further. Indeed, in February 1970, just two months before the hearings at the Piccadilly Hotel, a further £1 million development scheme was unveiled. If LA3 doesn't get you, warned the *Bucks Advertiser,* Luton will.

Roskill, of course, had to make a decision on the third airport without the benefit of knowing what the fate of that application would be. He assumed in his Report that substantial development was unlikely to be permitted; and in July 1971 this suspicion was apparently confirmed when the Government announced that Luton was not seen as a major airport, that discussions would take place with the Corporation, and that there would be stricter restraints on noise.[5] But at the time that the Roskill Report was compiled, Luton too was one of the great unknowns.

Then there was Southend. That, clearly, was going to be crucial in the Roskill equation. Among the vast wealth of accumulated benefits which south-east Essex would gain from the choice of Foulness (in the almost unanimous opinion of people living in the vicinity of Cublington, Nuthampstead and Thurleigh) would be the shutting of Southend Airport. And so inevitably the Commission got sucked into arguments about that.

Southend was a highly successful operation too. In 1951 the airport had handled 10,000 passengers and 170 tons of freight. In 1955-6 its three grass runways were replaced by two concrete ones and Southend promptly took off. Just under a quarter of a million passengers used it in 1959-60; four years later the total was just under half a million. By 1969 it was the fifth busiest airport for aircraft movements in the country, nosing out Luton which had once been well ahead of it.

In freight carriage its development was even more striking. In 1968 it was the second busiest freight airport in Britain, ahead even of Manchester and doing twice as much business (measured in quantity of goods conveyed) as Gatwick. It made profits of £100,000 or more in every year but one between 1963 and 1968.

Southend too had been considered as a possible third London airport. It had been favoured by BEA at the time of the Stansted inquiry, but was rejected by the Government partly because road access was considered inadequate and partly because the nearby artillery range at Shoeburyness was bound to curb its development. Like Luton, it was destined to run down if Stansted was built.

But Southend's future no longer seemed as certain as it had been. The passenger peak of 1966 was followed by decline; then a levelling out, but not at the 1966 level. Between 1968 and 1969 freight use declined by thirteen per cent; then it steadied, but there was a further ten per cent decline in 1971. Because of falling revenue, a ban on night flying was removed in 1969.

That made it an issue for debate at Roskill. How much alleviation of Southend's sufferings if Foulness were chosen was to be entered in the scorebook? Quite a lot, said those who were defending Cublington. Rather less, argued anti-Foulness interests, since it was clearly a declining asset and so, hopefully, a declining source of noise.

And then there was the whole question of airports outside south-east England; not, as the Roskill Commission insisted, part of their brief – yet incontrovertibly, as Buchanan suggested,[6] an essential element in a right decision. If Foulness were chosen, Birmingham and Manchester could be expected to grow faster than if the choice fell on Cublington. Was this, in terms of national policy, a good or a bad thing? Was it justified because of the stimulus it would bring to the economy of these regions? Or was it undesirable on the grounds of amenity? This issue, having been put on one side by Roskill, was to become a central preoccupation of those who later sought to overturn the government decision to go to Foulness. All this would have complicated any inquiry; but for one built, like Roskill, on the concept of weighing and balancing, the accumulated uncertainties caused by the absence of any coherent national plan created a hazard which must at times have been deeply frustrating.

Many of these uncertainties surrounding the Roskill Commission's work could be traced to the enforced flight from planning in the late forties and the substitution of assumptions for decisions which occurred in the fifties. But there were other factors too which helped to bring Stansted back into the running as one of London's major airports.

One of these was the elimination of Blackbushe. Blackbushe, it may be remembered, was one of the chain of seven airports which served London at the end of the forties. In March 1949 it was ruled out, along with Gatwick, and the decision to de-requisition it by the end of 1950 was announced. Then in 1953 it reappeared on the short list for the second London airport and became a 'supplementary' choice, standing in for Gatwick when necessary.

The A30 road, which by running across it had helped to rule it out in 1949, ran across it still and was getting busier. There was air control conflict with Heathrow, Farnborough and an RAF aerodrome which the 1953 White Paper was too shy to name. But it was still a 'good aerodrome', apparently with a lot of use ahead of it.

Yet by the end of the decade it was ready to close. It wasn't just the air control factor which closed it, though that was clearly important. The Ministry had concluded that it would not attract enough traffic to justify the compulsory purchase and capital expenditure of £2 million which was needed to bring it up to the right standard for bigger aircraft; especially when the activities which would then be made possible might create problems with Heathrow.

Blackbushe was still open for minor flying at the end of the sixties. It was run by the man who had taken the proving flight out of Heathrow in January 1946, Air Vice-Marshal Don Bennett – now far removed from the Liberal Party he had briefly represented in Parliament and active in the Political Freedom Movement with its dedicated sense of white Commonwealth solidarity and deep hostility to entering Europe.

Bennett continued optimistically to see busier days ahead at Blackbushe. But there seemed little possibility that the 1960 sentence would be repealed. There was, though, one minor consolation. At least no one had yielded to the advice of one Conservative back-bench aviation specialist who back in 1952 had been urging the sluggish and reluctant Ministers in Civil Aviation to get a monorail built to Blackbushe. There have been some unhappy moments in airports policy since 1945, but at least the building of Britain's first redundant monorail was never numbered among them.

5

... In which some awkward questions are asked at Stansted, Essex; and an assumption becomes a decision

And so it came about that on a May morning in 1961, a party of MPs arrived at the airport at Stansted in Essex to make a conducted tour; and, having completed that tour, sat down at a table with some of the aviation experts of the Ministry and proceeded to ask some exceptionally awkward questions.

The MPs were members of the Select Committee on Estimates and they were carrying out an inquiry into London's airports. They were led by Robert Carr, the Conservative member for Mitcham, who three years earlier had given up his post as a junior minister to attend to the family business. His release from ministerial duties had at least given him time to chair the sub-committee of the Estimates Committee which had been charged with this particular investigation. With him at Stansted were four other Conservative MPs – Sir Henry D'Avigdor-Goldsmid, Robert Gresham Cooke, Commander Stephen Lynch Maydon and Robin Turton, with one Labour member, Bruce Millan. Facing them were an Assistant-Secretary from the Aerodromes Planning Division of the Ministry of Aviation, Graham Hill; the General Manager of London Airports, Robert Edwards, and his deputy; the Deputy-Director of the Aerodromes Technical Division of the Ministry of Aviation, F. W. Hancock; the Stansted Commandant, J. V. Noyes; and the Ministry's Southern Division Controller, R. W. N. B. Gilling.

Stansted at this stage was a decidedly minor aerodrome. Built during the Second World War, it had been in use by the Americans until 1946, when it passed to the Air Ministry –

although the USAF retained an interest in it. It was taken from the Ministry in 1949: 'My department,' said George Lindgren in a written answer, 'is proposing to take over Stansted Airport as the principal short haul, bad weather alternate to London Airport and Northolt, and if practicable, facilities will be made available there for charter operators.' Subsequently, announcing the decision to release Gatwick and Blackbushe, Lindgren had promised that room would be found at Stansted for charter operators whom this decision would displace. Stansted, in other words, now rated number three to London Airport and Northolt; and Northolt's future was already distinctly in doubt, as Lindgren told the Commons in the same March 1949 debate:

I think all those who have given thought and study to the aerodrome pattern in London agree that the proximity of London Airport to Northolt makes the use of Northolt when – and only when – London is becoming used to the greatest capacity, a dangerous operation, or a risky one if not a dangerous one.

Two weeks later, Lindgren was able to tell the local member of Parliament, R. A. Butler, that Stansted was to be opened up 'progressively' to charter operators from as soon as 1 April. Families who had moved into the hutted accommodation on the airfield were going to have to be moved out, because the huts would be needed by people employed in airport work.

But the defence objection, together with the discovery that Stansted was on the wrong side of London and rather a long way out of it, had overthrown all that, and the 1953 White Paper which had picked Gatwick and Blackbushe off the scrapheap had pushed Stansted back towards it. The intention was that it should be leased out to anyone who thought it worth flying from there.

The USAF had not lost interest in Stansted however, despite its fall from grace in the eyes of the Ministry of Civil Aviation. In August 1956, *The Times* reported that American army engineers had been working 'night and day' on a £15 million scheme to convert the airfield for use by the heavy modern aircraft of

Strategic Air Command. 'When the work is completed,' enthused *The Times*, 'the Essex airport will be one of the finest air centres in the country, comparable with London Airport.'

The plan to make it an important base for Strategic Air Command very soon fell through, and the Americans once more departed. But the runway they had built – certainly one of the best in Europe – remained. Delighted as they were with this windfall, however, the Ministry of Aviation could not look with much enthusiasm on the state of Stansted as it then stood. The civil operators whom they had hoped to attract had not succumbed to its rather inadequate blandishments. In 1959–60 the total number of passengers using Stansted fell to 13,077; it recovered in the two following years, but this was almost entirely due to trooping contracts. And the place was running at a quite unacceptable loss. The closure of Blackbushe had caused the Ministry to have second thoughts about Stansted; the obvious assumption was that it must now be a contender for the number three place in the list of London airports. But for the moment, the obvious course was to put Stansted into cold storage, on a 'care and maintenance' basis. The airlines were not happy about it; BEA complained bitterly that they would be denied its use for training. But the losses involved in keeping it open – the figure for the previous year had been £200,000 – could not be ignored.

The losses, of course, were one of the reasons which had brought the parliamentary Sub-Committee to Stansted. As its name suggests, the Estimates Committee was largely concerned to discover what was being spent where, and why, and to point out ways of spending less of it to more productive effect.

This brief had guaranteed that they would spend much of their time investigating the state of Gatwick, which, some seven years after it had been picked out by the 1953 White Paper, did not appear to be doing very well. The financial side of the picture was extremely difficult to understand, and the Estimates Committee was to make recommendations for a more easily penetrable way of accounting. But that was not the only trouble. The longer standing mystery of what Gatwick was supposed to be

5

for, which had caused so much speculation in 1953–4, seemed no nearer solution in 1960–61.

It had originally been announced as a place to take diversions from fogbound Heathrow. Well, on that count, its life had certainly proved to be anything but action packed. Only seventy-six flights had been diverted there in 1958, 266 in 1959 and 141 in 1960.

It was also meant – though to what extent had never really been established – to attract operators displaced from Croydon and to take some of the weight off London Airport at the seasonal peaks. But the airlines did not seem to be very taken with it. The British national airlines felt that their proper place was at the main London airport; and foreign operators did not want to take second best. Air France was quite succinct in its objection: Gatwick, it said, 'isn't London'.

But if the Ministry was in two minds about the role it saw for Gatwick, the airline operators could be pardoned for being shy themselves. The Estimates Sub-Committee clearly felt the two problems were related and it tried manfully to sort them out. But the answers were less than definitive.

Was Gatwick a major airport – or merely a supplementary airport? Henry Hardman, Deputy-Secretary at the Ministry, didn't directly answer that; but he said it would be 'sizeable' – with a million passengers by 1961–2 (the figure for 1960–61 was less than half a million) and perhaps two and a half million by 1970.[1] But wasn't that, Robert Carr persisted, a contradiction of the soothing statements in the 1953 White Paper? 'I think it is perfectly true,' replied George Morris, Under-Secretary in charge of the Aerodromes Planning Division, 'that assurances were given to the public inquiry that Gatwick would not become a London airport in the full sense of the word, which is rather different from saying it will not be used to capacity.'

There may indeed be a difference there, though it is one that the jet-haunted people of Charlwood and Lowfield Heath might find it difficult to appreciate, and it certainly did not satisfy Robert Carr and his colleagues. Wasn't Gatwick, Carr suggested to the Airports General-Manager, Robert Edwards, falling between two stools: too good to be a mere diversionary airport, but not good enough to attract regular scheduled services flown by the biggest and most modern aircraft? 'In my view,' said Edwards, 'we should be bold and press on with the development of Gatwick, even though we are accused at the moment of having something like a white elephant.'

Between the lines of the Ministry's answers a clearer scenario seemed to emerge, though imprisoned as they were in the pledges of Ministers nearly a decade earlier, it was perhaps difficult for them to spell it out in detail. But the expectation was that the gradual clogging up of London Airport would gradually convince the airlines of the merits of using Gatwick, and development would proceed from there. It was, however, an assumption, rather than a decision, and the Committee wanted a decision. Insufficient attention, it declared in its subsequent Report, had been paid to specifying exactly what role Gatwick was intended to play. The Ministry must clarify it now.

The long-running saga of 'vacillation and indecision' (in the words of the Committee Report) at Gatwick, which had opened in the late forties and clearly showed no signs of closing yet, heightened the Sub-Committee's anxieties about Stansted. Here was an airport placed on care and maintenance on the assumption (not decision) that it would have a more glorious role in years to come. 'It would be rather frustrating,' observed Robert Carr to one official witness, 'for the taxpayer to have to spend half a million over the next five years [the estimated cost of care and maintenance was £100,000 annually] for the airport to be then given up.' But the evidence heard first in London and then at Stansted itself did very little to convince the MPs that there was not at Stansted another Gatwick saga in the making.

The 1960–61 Committee's evidence and report, indeed, constitute a remarkable and very readable document which catches the Ministry's experts as if in a tableau; frozen, as it were, in postures of indecision and inactivity. Yet at the same time, it does much to explain how this frozen condition had come about.

That a third London airport was going to be needed had long been an article of faith. (It involved the admission, incidentally, that Gatwick was indeed the second London airport, and Ministry witnesses to the Committee did not avoid making statements which clearly identified it as such.) A committee under Sir Eric Millbourn, industrialist and company director, had been set up to consider the development of London's airports in October 1955 and reported in April 1957. This, unlike the 1953 White Paper, which had dealt only in generalities, produced forecasts of future traffic and transport movement which suggested that in 1970 Heathrow would be working at full capacity (this was estimated at eighty movements an hour at peak and a sustained rate of sixty-four). They also pointed out that the tide of passengers would swell much faster, since the aircraft moving on the tarmac by the end of the sixties were likely to be much bigger than those in use at that time. Gatwick, they concluded, would require further development to take the overflow and a third London airport 'might' be needed.

Fate, therefore, was pointing its wavering forefinger at Stansted from this time onwards (Blackbushe now being on the downward slope). But while Ministry officials were by 1960 convinced that a third airport was becoming inevitable – and while subordinate decisions, such as those on the establishment of military bases, were for some time being made on the *assumption* that Stansted was the obvious site – the final decision was not regarded as urgent.

Having disposed of the courteous preliminaries at Stansted, Robert Carr and his colleagues began pressing the Ministry witnesses as to their intentions for the airport. Their reply made it clear that the purely fortuitous appearance of the mammoth runway had done much to raise Stansted's chances in the Ministry's estimation. Graham Hill summed up the position like this:

It has a runway . . . it is a very good runway, and we own it. There is no real problem of dispossessing people, going through lengthy public inquiries and the rest. It is a perfectly workable airport for the future. The general position is that as far as we can see, Heathrow and Gatwick between them should be capable of development to handle the bulk of the London area traffic for the rest of this decade. After that, as was foreseen in the 1953 White Paper, yet another airport might be needed – Stansted can certainly play a role here.

We cannot say for certain whether Stansted will be either *the* third London airport or *a* third London airport, but we can most certainly say that it would be imprudent to get rid of it now, because we can foresee definite uses for it in the future, going right up to the role of *the* third London airport.

But what, the Committee wondered, had happened to the factors that had made Stansted unsuitable in 1953? Had the defence objection disappeared? Not entirely, it emerged. George Morris had admitted at one of the Sub-Committee's earlier hearings that military flying at Wethersfield might be a problem: 'however', he added, in one of the most significant single under-statements in the whole history of airports policy, 'military plans do change.'

Hadn't there also been worries about possible air traffic con-flicts if the present runways were used? Yes, there was some truth in that too. You would need to acquire land to get your necessary additional east-west runway, admitted a technical witness at the Stansted hearing – thus somewhat undermining Hill's contention that problems of dispossession and lengthy public inquiries might be avoided.

The MPs seemed baffled by the apparent view of the witnesses that it was quite safe to let the fate of Stansted simmer for a while longer yet before deciding one way or another. That, they felt, with one eye on the cost of care and maintenance, might be an uneconomical solution; but strictly in terms of policy, it also seemed a dangerously permissive thing to do. 'What we are concerned with,' said Robert Carr at one point, 'is what it is you are doing to make yourselves as decisive about your future plans as quickly as possible. Are you as urgently as you can making up your minds decisively what the third airport of London shall be?' The answer was less than categorical: 'I sincerely hope so,' the Ministry witness replied.

It was clear that the Ministry's present intention was to defer its final decision until 1965. Why, Carr asked, would it be easier to decide then rather than now? 'There are a number of changing factors,' Robert Edwards replied. '1965 is the last date: we should have made up our minds by 1965 if we are going to be in time for the need.'

The sums were fairly simple. It was agreed that an airport was likely to be needed in the seventies. It was intended that the decision be made by 1965. That, conceivably, meant cramming the whole operation into five years – settling on a site, working out the consequential details, putting it before a local inquiry and waiting for the Minister's report, constructing it and getting it open before Heathrow and Gatwick got clogged solid. It might make sense to the Ministry but it made very little to the MPs. Could you really, they persisted, do a job like this in five years? The Ministry were satisfied you could. They had done it at Gatwick; the White Paper, Hill pointed out, had appeared in 1953 and the airport had been operating in 1958. But that didn't satisfy the Sub-Committee either. Gatwick, they recalled, had been built close to a main line and what by the standards of that time had been a good main road – the A23, whose diversion around the site had caused such sad consequences for its roadside cafés. That wasn't the case with Stansted. 'You said five years,' Robert Carr objected, 'but that's not long enough to deal with the roads, is it?' But it was too early to do anything about the roads at this stage, said Hill; you couldn't badger the Ministry of Transport now as you would if you were sure.

It was the fate of Thales, as set out by Ivor Thomas in the 1946 debate, which seemed to be preoccupying the Ministry. Hill explained it to the Sub-Committee:

I think the answer is the difference between seeing ahead and looking ahead. We are continually looking ahead as hard as we can, but we cannot see indefinitely ahead. You have to bear in mind that round the corner there are revolutionary possibilities – supersonic aircraft, short take-off and possibly vertical take-off types – which may have completely different ground requirements. Therefore we have to do everything we can now to get the information on future developments to make sure that we do not make a mistake in building a third airport somewhere that would have been very good in the sixties but will be no use in the seventies.

It may have sounded like the defensive attitude of someone who has been undergoing what, despite its courtesy, must have been a fairly unpleasant grilling; but in fact the answer exposes one of the central problems of the problems of governments in dealing with rapidly developing technology. There was to be an argument on precisely the same piece of territory after the Roskill Report, with some of those who would have most strongly condemned the Ministry's delaying tactics at the Stansted session of the Committee now arguing a somewhat parallel case. In fact, the calculation they made does seem, with the advantage of hindsight, to have been as tainted by complacency as the Estimates Sub-Committee clearly took it to be. The Stansted decision was indeed finally taken in 1965, to be followed, as should have been predictable, by public controversy, a bitterly contested local inquiry and a government re-examination and eventual decision, which occupied something like two years; and at that stage, the Ministry were convinced that the time for further discussion would have to be curtailed because of the extreme urgency of getting started. And the unavoidable necessity of making consequential decisions should also have pointed to the need for an early decision. But to anyone who was liable to get saddled with subsequent blame for building a good sixties airport which withered into redundancy in the seventies, keeping one eye open to the fate of Thales did not seem an imprudent precaution.

The Committee's eventual judgment, however, was stern:

> It is important that a decision should be reached in the com-
> paratively near future whether a third London airport will
> be needed within the next fifteen years and if so whether
> Stansted is the best site. Your Committee are anxious that the
> vacillation and indecision which marked the early history of
> the development of Gatwick should not be repeated. It would
> be a serious matter if it was ultimately decided not to pro-
> ceed with the development of Stansted after large sums of
> money had been spent either on keeping the airport open for
> training and charter flights or on putting it on care and
> maintenance.

That did not mean that they favoured an immediate decision
in Stansted's favour. They were not equipped to decide whether
or not this was the best site; they had considered no other, and
had not heard enough relevant evidence. Indeed, as far as they
could reach an opinion it was a sceptical one:

> Your Committee have not received evidence about other
> possible sites for a third airport for London, but they consider
> that Stansted has certain serious disadvantages as a potential
> London airport.
> Even if a rail link with London could be provided, some
> major replanning of the road to London would have to be
> undertaken in co-operation with the Ministry of Transport
> . . . In spite of the undeniable difficulties of forecasting future
> airport requirements for London, your Committee consider
> that this period of suspended judgment cannot be permitted
> to continue for long.

The Ministry of Aviation, with the Air Ministry and Ministry
of Transport, should carry out an immediate and detailed study
of the prospects of Stansted as a future third airport for London,
and the announcement of a firm decision should be regarded as a
matter of urgency.

And so the 1960–61 Select Committee passes out of the story, having carried out faithfully the invaluable role of committees of this kind, which is to say, digging its nose into other people's business and pointing out to the Minister things about his Department's policy which he has failed to deal with himself. The Minister who had to reply to the Report was Peter Thorneycroft, who had succeeded Duncan Sandys in July 1960 and who would, in the traditional course of the repeated ministerial reshuffling which all British Prime Ministers favour, give way to Julian Amery in July the following year.[2] His response was in line with what the Sub-Committee recommended. 'As for Stansted,' he told the Commons on 28 May 1962, 'we are studying this hard now. We are considering with the airlines and with the Ministry of Transport, because it is very much concerned, and with the airline associations and with others, whether Stansted will prove to be the most suitable aerodrome as the third airport for London.' The form in which these disparate interests came together was an inter-departmental committee.

The inter-departmental committee is a routine Whitehall device for dealing with those situations, of which there are inevitably many – even now, in these days of fewer and bigger Ministries – where several different Ministries have a finger in the same governmental pie.

Their membership may not necessarily be confined to civil servants. Instead, as is the normal practice with departmental committees, representatives of outside interests which have become heavily intertwined in government decision-making may be invited to join the committee. There is an obvious advantage in this: they will be able to point out any proposal which they find to be intrinsically unworkable; their presence should also guarantee that none of what the committee finally produces is so contentious as to cause grave political inconvenience later on. Some of these committees are called into being – as the Stansted one was – to deal with one specific situation; on other occasions, they will deal with continuing problems, and may be in existence for years.

The kind of report which a committee eventually produces,

however, will inevitably be governed by the composition of the group which sits down around the table. One department may be heavily represented, and another may be represented only thinly – or even excluded altogether. What is more, the outside interests who are chosen to sit on the committee may be able to deal adequately with one aspect of the problem, but may leave out altogether other considerations which are believed – perhaps quite wrongly – to be of less critical importance.

The Inter-Departmental Committee on the third London airport, between 1961–3, demonstrates the classic dangers in staffing a committee. Of its fifteen members, thirteen were from aviation. The Chairman was George Vincer Hole, Under-Secretary in the Ministry of Aviation's Aerodromes (General) Division. He had with him seven other Ministry of Aviation civil servants; two representatives of The National Traffic Control Services – one of them, Captain Vernon Hunt who had given evidence at the Gatwick inquiry; two representatives of the state airlines and one from the British Independent Air Transport Association. The remaining two places were filled by an Assistant-Secretary from the Ministry of Housing and Local Government and by a man from the Ministry of Transport. That meant that non-aviation interests in Whitehall had been awarded only two places, and non-aviation interests from outside had been awarded no places at all.

The most striking thing about the Committee, looking at it now, is that the Ministry of Housing should have been so meagrely represented. Indeed, the relationship between the Ministry of Aviation and the Ministry of Housing (the department principally responsible for planning policy) in these years, reminds one irresistibly of two men carrying a heavy piece of furniture. The one at the front – which in this case is the Ministry of Aviation – is making decisions and giving instructions; the one at the back – the Ministry of Housing – is simply responding as and when necessary. It is the Ministry of Housing which has to set up the local inquiries, like the one at Gatwick and the one which was later to be held into Stansted. But of any assertion of the rights of the Ministry of Housing to take a substantial, let alone a leading, part in policies of airport location, there appears at this time to be very little sign.

Perhaps this would have been more vigorously pointed out in 1961 had the composition of the Inter-Departmental Committee been generally known at that time. (The significance of Thorneycroft's statement, in which the Ministry of Housing was not even mentioned, appears to have gone unremarked.) But it is not the normal practice to identify members of these committees. Civil servants are expected to work in decent anonymity and the subsequent naming of this Committee was an exceptional course.

Inevitably, by assigning this task to this group of people, the Government had ensured that the eventual report would deal with the decision largely in terms of aviation suitability. Had the Committee been designed purely to serve as consultants on the aviation factors involved, of course, that would have been fair enough. And indeed, there were some attempts later on to suggest that this was what had been intended. In the run-up to the 1964 election, R. A. Butler was having a little local difficulty in his Saffron Walden constituency over the proposal to expand Stansted. Six weeks before election day Butler wrote to Sir Roger Hawkey, one of the leaders of the North-West Essex and East Hertfordshire Preservation Association, which had become the leading popular opponent of the airport: 'So far, an expert inter-departmental committee has recommended Stansted Airport on technical grounds. No account has yet been taken of the human considerations, the living conditions, the amenities and the agricultural character of the area involved.' Two days later Butler was able to publish a letter he had had from the Prime Minister, Sir Alec Douglas-Home, which put the Stansted recommendation in exactly the same light.

The Committee's recommendations in favour of Stansted [wrote Sir Alec] were derived from a study of the operational and other technical requirements of an airport. The Government recognises that whilst these technical considerations carry weight, many other considerations arise such as the disturbance of local life in villages and towns in the neighbourhood, loss of amenities and loss of good agricultural land. The Government has in no way made its final decision and cannot do so until it receives the report of the inquiry.

Very reassuring for people in the Stansted area, of course – if only they could have brought themselves to believe that this really was the truth about the Government's attitude to the Inter-Departmental Committee's Report. But it hardly squared with the much more robust assertions of Julian Amery in his foreword to the Hole Committee Report:

The report ... concludes that Stansted should be selected and designated as London's third airport. The Government believe that this is the right choice.

Every effort will be made, in conjunction with the national and local authorities responsible, to ensure that the airport and its surface communications are developed to the best advantage of the travelling public. In particular, we shall consider closely the implications of this decision for communications between Stansted and central London and between the three London airports. We shall also seek in accordance with the recommendations of the Wilson Committee on the Problem of Noise to reduce the impact of noise on residential areas near the airport.

I hope that there will be full public discussion of this report and I shall welcome constructive suggestions for making Stansted an efficient and attractive airport.

And, if the Government – as Sir Alec and Butler had indicated – accepted that there was still much more work to be done on the planning implications before an aviation recommendation became a final decision, then there was precious little indication that anyone was getting on with it. Indeed, as will be seen, the witnesses who appeared for the Government at the Stansted inquiry were nearly all drawn from aviation too. The Inspector was asked to inquire into the suitability of Stansted for the third airport; yet the Government managed to produce for his assistance not one qualified witness on the planning issues; indeed the opposition to the proposal, who were very much aware of the planning consequences of the decision, were given more or less a free run.

The Committee met eleven times between their appointment on 21 November 1961 and the completion of their work on 23 May 1963. The criteria which they chose for a potential third airport rather undermined the subsequent assertions that their brief had been purely concerned with aviation. The new airport, they said, should interfere 'as little as possible' with agricultural, local industry and the existing amenities of the neighbourhood. As for noise, 'the airport should be sited so as to disturb as few people as possible'. The Committee took the view that Heathrow and Gatwick would be able to cope with incoming traffic longer than the earlier Millbourn Report had suggested. But the need was nevertheless becoming urgent. By 1972, the existing airports might be having to turn traffic away. Even a second runway at Gatwick, which they advocated, would not be enough to cope. So a third airport would certainly be needed in the early seventies.

The next question was where it should be put. They started with a list of a hundred sites which they quickly whittled down to eighteen. Several sectors were eliminated altogether. The area south and south-east of London, and the area to the north-west, were all ruled out because of the difficulty of routeing traffic there. (Cublington would never have got past the Inter-Departmental Committee.) Sites west of London would have to be well away from the capital because of the difficulties of interference with operations at Heathrow. The last eighteen included Luton, Southend and Stansted, Hurn – the stand-in London airport of fifteen years before – and coastal sites at Cliffe and Sheppey in Kent and Foulness in Essex. It was admitted quite casually in the Committee's Report that only one of these eighteen sites – Stansted – had been visited.

What was more, the insistence that the chosen site should be no more than an hour's journey from London effectively ruled out virtually all the eighteen sites which they were supposed to be considering. It certainly removed all the western sites and it told heavily against most of the eastern ones also. Then there were the objections of the Ministry of Defence. The evidence of the Ministry was that several of the sites being considered would interfere with the operations of the artillery range at Shoebury-ness; and that could not be allowed. 'Shoeburyness', said the Report 'is considered by the War Office as irreplaceable because it is unique in possessing a large area of hard, flat sand, which enables shells to be recovered intact for examination.' The coastal site at Sheppey, which in some ways came out quite well from the analysis, would conflict, said the Committee, both with Shoe-buryness and with the airport at Southend, which would be forced to shut. Foulness was 'suitably flat and sparsely inhabited and aircraft might be able to take off and land over the sea', but unfortunately on other grounds it was quite unrealistic. It was difficult to get to by road. It was incompatible with the continuance of Southend Airport. And it would cause serious conflict with Shoeburyness next door.

And so, in the end, the site which had been in the Ministry's mind since the fifties (and which had been the only one considered worth visiting by the Committee) turned out to be the best one. Just as Heathrow had been, nearly twenty years before 'the only possible site on which a great airport for London could be built', so Stansted was now 'the only one with a clear prospect of success'. It was, for one thing, the only one on the list which satisfied the criterion of being within one hour's travelling time of the West End. It had the advantage that much of it was al-ready in use as an airport – and of course, it had the runway which the Americans had so happily provided. Admittedly there was an air traffic disadvantage in that, if the airport were developed to operate with parallel runways, a conflict with Heathrow might occur from time to time; but this could be avoided by different runway orientation at Stansted, though at the risk of creating a noise problem in Bishop's Stortford, Great Dunmow and Stansted itself. By the time they had finished rounding up the advantages and disadvantages, the Committee sounded positively euphoric about what the airport might do for this area of Essex.

The airport, together with the new motorway, would greatly enhance Stansted's attractiveness as an area for the development of commerce and industry and would make the locality a strong candidate for planned increase in population on a larger scale. This in turn would greatly benefit the airport. The selection of Stansted would provide an excellent opportunity for the co-ordinated planning of the airport and nearby residential and industrial zones so as to avoid the noise problems that have created so much difficulty at Heathrow.

The Committee added that its recommendations were unanimous.

Though the Report was submitted in June 1963 it was a further nine months before it finally appeared, with Julian Amery's encomium attached to it. There was no statement in the Commons, and the publication of the Report did not create any immediate public excitement. Clearly there would be no time now to act before the general election, which was now not very far away.

When the election came, R. A. Butler survived with no difficulty at Saffron Walden (an Independent who had planned to stand against him on the airport issue thought better of it). But the defeat of the Conservatives ensured that a new Minister would now have to preside over the political consequence of the recommendation which the Inter-Departmental Committee had produced. The responsibility of setting up the Inter-Departmental Committee, whose work came to be the target of such sustained and penetrating criticism by the airport protestors and the experts advising them, belonged to the Conservatives; the responsibility for standing by the work of that committee, long after its inadequacy had been unanswerably demonstrated, was to fall on Labour.

6

. . . In which, Julian Amery having declared that Stansted is right, Douglas Jay defends it against all comers

What finally set the cauldron bubbling at Stansted was a meeting of local authorities at Harlow on 16 July 1964 when the detailed implications of what the Inter-Departmental Committee had recommended were spelled out for the first time. 'A giant runway, more than three miles long and pointing directly at Harlow is planned for the new Stansted Airport' reported the *Harlow Gazette*, which had got to see and photograph the plans, on 24 July. When the local authority representatives who took part in the Harlow meeting got back to their council chambers, there was consternation and anger. The local press reported a 'storm' at Dunmow Rural Council on Monday 20th as the proposals were unveiled, and a week later there were reports that Sawbridgeworth, close to Bishop's Stortford, foresaw 'terrible effects'. An area of nineteen square miles, it was deduced, was directly threatened. A meeting called by Mr and Mrs Edward Judson, who lived at Great Dunmow, to discuss the revelation of these plans led to the formation the following month of a North-West Essex Preservation Association. Later, with recruits from across the border in Hertfordshire (the country boundary runs within about a mile of Stansted) it became the North-West Essex and East Hertfordshire Preservation Association.

This organisation was to recruit some 13,000 members and to raise nearly £25,000 in time for the inquiry. But it was by no means clear that they spoke for everyone. One of the most fascinating and intricate speculations for those who followed the Stansted story was whether or not there was really a formidable

6

body of opinion which badly wanted the airport to come to this part of Essex. Events began to be depicted in terms of class struggle. One BBC television programme, for instance, described the mouth-watering prospects of a battle between the property owners and the workers – an over-simplification which brought protests from the anti-airport campaigners and led the BBC News Editor to agree that the report had been 'unhappy'.

The whole of this controversy has some historical importance because of the use which came to be made by the Labour Government later on of the argument that the protests were a largely middle-class affair, while the voice of the less articulate working-class people of the area was going almost unheard. The grass-roots equivalent of the NWEEHPA was an organisation, launched (mainly by workers at the existing Stansted Airport) at a meeting at the Barley Mow public house in Stansted on 20 February 1965 called the Stansted Area Progress Association. Fred Browne, who was elected Chairman at that meeting was a local organiser of the National Union of Vehicle Builders at the airport. The Association later produced before the Stansted inquiry 927 letters from local residents and workers who did not object to the airport proposals, subject to reservations on compensation and the preservation of amenities. The Association quickly became involved in direct conflict with the NWEEHPA, particularly in the columns of the local newspapers. The war was conducted with an undisguised bitterness. SAPA accused the Preservation Association of being dominated by well-to-do people. It said too that local district and parish councils in the area had reflected the feelings of that same social group rather than of the whole of the people they were supposed to represent. In a submission for the Stansted inquiry SAPA declared that the Rural Councils of Dunmow and Saffron Walden and the Borough Council of Saffron Walden were not accurately representing the feelings in their districts. 'The preponderance of members representing the country land owners, farmers, property owners, business and commerce, not to mention the future developer, both genuine and speculative, outweigh the ordinary industrial and agricultural workers' and weekly wage earners' interests.' There were even allegations of skulduggery and fixing. Of the parish meeting held at Stansted on 26 October 1965 they said:

> Our Association think it would be of interest for the Inspector to note that shortly after this meeting had commenced the local fire alarm sounded, necessitating a number of the public who were present and who were members of the fire service – professional and part-time fire fighters – having to leave the meeting. It was known that they would all have voted in favour of the airport or against a rate contribution to the Stansted Preservation Association.

It was also true that the front of councils opposing the airport was less monolithic than outsiders sometimes assumed. Two of the councils crucially involved who appeared as objectors at the inquiry did so only after considerable delay and doubt. In February 1965 the Hertfordshire County Council planning committee agreed to recommend to the County Council that the proposals for Stansted would be acceptable if the proposed runway plan was modified. They said the county's case against an airport would not be a strong one. Despite vigorous harrying from the Councillor for Sawbridgeworth, Miss L. A. M. Lloyd-Taylor, the County Council continued to delay a decision. At the end of May, when Miss Lloyd-Taylor exhorted her Council to follow the example of Essex, which had voted seventy-six to five in a poll to oppose the Government, she was told that 'considerable study' was still necessary before a decision could be reached. At the end of July, Hertfordshire were still undecided and 'awaiting more information'; the planning committee now declared that since the site of the proposed airport was not in the county, it could not object in its role as a local planning authority. The eventual decision to oppose the airport, which was unanimous, was not taken until 30 November 1965 – only six days before the inquiry at Chelmsford was due to open.

At Bishop's Stortford, a town of 21,000 people only three miles away from the airport, which, it was feared, might be heavily affected by noise, there was also a long period of indecision about what attitude to take to the airport development. Clearly there were those in the town, particularly among the shopkeepers, who were strongly attracted by the Government's proposals. A succession of monthly meetings failed to bring the Council off the fence. Even as late as November 1965 there was still some doubt; the Council's noise consultant, who subsequently gave technical evidence at the inquiry, had advised them that they had little case for an objection on noise grounds. In the end it was unanimously resolved that they should appear at the inquiry; but even that did not mean total opposition. The Councillor who proposed the resolution was actually an airport supporter – he believed that economic benefits would outweigh disadvantages; but he pointed out that the only way for the Council to be able to give its views on noise and traffic issues was to make an objection, since otherwise it would have no right to be represented at the inquiry.

Even Stansted village itself was by no means totally opposed to the airport; a parish meeting on Wednesday 13 October 1965 voted ninety-eight to sixty-eight on a motion to reject proposals for a major international airport at Stansted. A second resolution calling on the parish council to support the fight by donating £100 to the Stansted Preservation Association was carried by only six votes. There was then a demand for a parish poll, and this was held on 23 October. But only twenty-nine per cent of those eligible bothered to take part. The resolution opposing the airport was then carried by 434 to 244 and the one sanctioning the vote of money was carried by 351 to 288. The class bitterness which had appeared in many of the SAPA representations was also apparent here.

Miss N. Gillam [reported the *Herts and Essex Observer* later] said she would like to know how many people who voted against the airport in the poll have lived in Stansted for more than seven years. 'The airport was built during the war,' she went on, 'and it has been on the books that it was going to be a large airport for the last seven years. If the people who moved into the village between 1958 and 1965 did not realise this, then all I can say is that they have not read their newspapers, listened to their wireless or watched their television.'

The Labour-controlled local authority at Harlow (population 66,000), was clearly committed against the airport from an early stage. But the local Trades Council emerged – although very late in the day – as a supporter. In 1967, when the choice of Stansted was confirmed in spite of the critical findings of the Chelmsford inquiry, a meeting of the Trades Council, attended by eighteen delegates, passed a resolution favouring the airport. The meeting was then adjourned in order that the delegates could go straight to the Town Hall where a protest meeting against the airport, organised by the Town Council, was taking place. They attempted to put forward the resolution they had passed as a statement but were told it could only be put forward in the form of a resolution; which was done, and it was heavily defeated.

So it was certainly possible to say afterwards – as Ministers repeatedly did in the summer and autumn of 1967, and as some were still asserting five years afterwards – that the airport had fervent support in the area where they were proposing to put it. Yet, both among the local authorities and among such men and women in the street as felt strongly enough to take sides either way, antagonism towards the airport was far stronger than support for it. To say that opinion was divided in and around Stansted is true; but to go further, and declare that 'people wanted it to be there' was a dangerous over-simplification, and one for which those who espoused it would have to pay in the end.

The leadership of the Preservation Association was – as its critics said – overwhelmingly middle class. That was, no doubt, unfortunate. But it was also in the nature of things that some of the articulate middle class (and indeed upper class) in an area of this kind would find themselves in the driving seat. Among the 13,000 members of the Preservation Association, among the passionate letter-writers to the local press, among the people

who turned up at parish councils and spoke passionately against
the airport and voted money for the campaign against it, there
were certainly people whose working-class pedigree could not be
challenged. It was difficult too, to challenge the socialist creden-
tials of the Harlow Council, or of the regional councils of the
TUC and Labour Party, both of which strongly opposed the
airport, advocating instead a site in the wastelands of Norfolk.

The inquiry into the Stansted Airport opened at Chelmsford
on 6 December 1965. It was a discretionary, not a statutory inquiry; that is to say, it was offered by the Government as an act
of grace, not forced upon it by the law of the land. This distinction would become crucial later on. The new Labour Government had repeated the promises of Sir Alec and R. A. Butler
that the airport question was still wide open and the inquiry
would be free to take whatever attitude it wanted to the proposal.
    Indeed, although this was not apparent at the time, the new
Minister of Aviation was himself very dubious about the wisdom
of the choice. Roy Jenkins had been a surprise choice as Minister
of Aviation. It was not a field for which he had exhibited any
outstanding public passion and it had rather been expected that he
would be marked out for a job in the economic field. But in fact,
Jenkins took on this assignment with enthusiasm and indeed in
the reshuffle after the Foreign Secretary lost his by-election at
Leyton in January 1965, he turned down a chance to forsake it in
favour of the Department of Education. That autumn, Jenkins
treated himself to a conducted tour of Stansted and some of the
alternative sites which the Ministry regarded with so much dis-
favour. This strongly confirmed his existing doubts about the
choice of Stansted. He found in the Stansted landscape a delectable
charm, a sunny contentment, which none of its rivals as a potential home for jumbo jets could match. Indeed, it is distinctly
possible that the whole case might have been re-opened at that
stage had an inquiry not been pending.[1]
    However, if his private preferences were still unknown, his

reputation as a civilised and rational politician was firmly established. Thus, as soon as his appointment became known, he
was besieged by requests to meet delegations. His response,
though basically non-committal, was sympathetic; and he went
out of his way to repeat and reinforce the pledges made about the
inquiry by his Conservative predecessors.

On 13 September 1965 the Chairman of the Essex planning committee, Brigadier T. F. J. Collins, led a deputation
which warned Jenkins of the suspicions which many of the
protestors attached to the inquiry. On 24 September Jenkins
wrote to him:

> it was represented to me that, unless the terms of reference
> covered the question of timing, it would be possible for
> Ministers to make the inquiry a formality by refusing to
> follow up indications in the Inspector's Report that an alter-
> native site was preferable to Stansted, on the pretext that
> the necessary study and survey of the alternative site would
> take too long and development of the alternative site could
> not wait. I assure you that this will not occur.
>
> If the outcome of this inquiry is that another site is to be
> preferred to Stansted, this will be followed up and it will
> not be ruled out for lack of time to study and survey the site.

This assurance was compounded by the choice of the Inspector
and Technical Assessor for the public inquiry. The Inspector was
not drawn from the regular team which is provided by the
Ministry of Housing for local inquiries. Instead the Government
appointed an unmistakeably independent figure in G. D. Blake,
a senior partner in a Kensington firm of estate agents. The
Technical Assessor was J. W. S. Brancker, a former official of the
International Air Transport Association, with long experience
of airlines and airports in Britain and North America.

The scale of the inquiry, however, was altogether smaller
than some had supposed when it was promised after the Inter-
Department Committee had reported. In spite of the enormous
implications of the scheme, the inquiry was much the same as

might have been set up to deal with an ordinary run-of-the-mill planning dispute. Its scope was further reduced by the terms of reference which, as was ever the case, were chosen by the Government, thus guaranteeing that the inquiry would be conducted on the Government's terms and not on terms equally acceptable to Government and Opposition. The Inspector was instructed 'to hear and report on local objections relating to the suitability of the choice of Stansted for an airport and the effect of the proposed development on local interests. It will be open to objectors to suggest modifications to the outline scheme of development, or to propose alternatives, but not to question the need to provide a third major airport to serve London.' Though the objectors baulked at the narrowness of these terms, it can be argued that they benefited in the end. Rigas Doganis, in his Fabian Tract, wrote: 'the inquiry had particularly local terms of reference . . . an inquiry at any of the possible sites with a similar reference, would inevitably have come to the same conclusion.' In other words – the choice of an airport anywhere demands a local sacrifice for the national good. By choosing such predominantly local terms of reference, according to Doganis, the Government had ensured that the sacrifice would be much more discussed than the national benefits which were to be achieved by it.

The inquiry lasted thirty-one days and heard 260 objectors, concentrating mainly on the change in the character of the area which the development of the airport would produce, on the noise and on the loss of agricultural land. The objectors, headed by Essex County Council and the NWEEHPA, had mustered a formidable body of expertise – one of the most impressive ever to be assembled at any such inquiry; and they were determined, however tight the terms of reference might be, to draw in those aspects of airport planning policy which the Government seemed to have left out of its calculations. This evidence – as subsequently expanded in a report of the Stansted Working Party in 1967 – produced telling comparisons between the depth of the investigation made by governments in other countries for planning major airports and the apparently restricted and somewhat superficial operation carried out by the Inter-Departmental Committee, with its built-in tendency to confirm a choice already in the Ministry's mind. The New York Port Authority reporting in May 1961 on airport requirements in the New Jersey-New York region had, they showed, included major research projects specially commissioned on the significance of plans for the regional economy, the economic effects of the airport, its effect on real estate values; estimates of future passenger plane, cargo and general aviation activity, airport design and engineering considerations; land use and ground transportation effects; the effects of accessibility on the volume of traffic using the airport; and the climatographical considerations.

On questions of regional planning – on which the Government side had called no witnesses at all – the Preservation Association produced Professor Lewis Keeble, then president of the Town Planning Institute, to testify that any large international airport – possibly twenty square miles in extent – must inevitably make a very large gap in the fabric of a densely populated country like England and would have dramatic effects on the economy of the region into which it was put. If the Government was seeking an expansion in this particular area, said Keeble, it certainly did not require the deliberate stimulus of an airport in order to achieve it. All you needed to do was to 'issue a planning permission and retire to a safe distance'.

In spite of the apparent prohibition of any general discussion of airport policy, the widening of the terms of reference since Gatwick also enabled the objectors to take that 'constructive' role which had been denied to the county councils who fought the Gatwick inquiry. They could, and did, suggest alternative sites. Essex County Council had come equipped with a plan to put the airport at Sheppey; while the Preservation Association had picked out Padworth, just south of the A4 London-Bath Road, some seven miles west of Reading in Berkshire – a suggestion which, apart from the not inconsiderable objection that it involved the closing of the Atomic Research Station at Aldermaston, about a mile away, and would mean suspension of flying at the Royal Aircraft Establishment at Farnborough, caused an understandable flurry of alarm when it was carried by the next day's newspapers into homes in and around Padworth.

The professionalism which the County Council and the Preservation Association had now begun to muster – the Preservation Association had become an altogether more sophisticated body since its first tentative and untutored days – ensured that a double attack was made on the Ministry's proposals. Firstly, there were all the objections made in human terms about the disruption to the lives of people who lived in and around the site – Mr and Mrs Miles of Sawbridgeworth were singled out by the Inspector for special praise for the way they depicted this aspect of the case. And secondly there were the wider grounds of national planning, which, since they had been so remarkably neglected by the official side, the objectors had more or less to themselves. But the Ministry case was made to look even worse by the lack of preparation which seemed to have been given to certain aspects of it – even though it was now four years since the Inter-Departmental Committee had begun its work.

This was dramatised early in the inquiry when Government witnesses came to the awkward question of the American base at Wethersfield. This, as had been admitted before the 1960–61 Select Committee, was likely to cause serious air space conflict with the new expanded Stansted. The Ministry, however, was ready with a perfect solution. Wethersfield, it was revealed by George Hole in evidence to the inquiry, was going to close. And when had this been decided? The decision, said Hole, had been taken at a meeting since the inquiry opened. Nothing about the need to close the base had been said to the USAF, he added in answer to a flurry of supplementary questions, until ten days before.

The disclosure of this startling piece of improvisation was followed by a damaging admission by another Government witness about the time it would take to get from central London to Stansted. This was, of course, an absolutely crucial consideration in the choice of Stansted; the Inter-Departmental Committee had laid great weight on the calculation that, alone of the sites considered, it met the test of being accessible to central London within an hour. But now a Ministry of Transport witness agreed with the contention of Counsel for the Preservation Association that the journey time from Stansted – by a new road, details of which had again not been revealed until the inquiry had started – would take between eighty-five and ninety-five minutes at off-peak times. Stansted, in other words, owed a substantial part of its selection to the fact that it was accessible within an hour; yet now it appeared that it was not accessible within an hour after all.

Blake's Report which arrived on the desk of Douglas Jay, President of the Board of Trade, who had now inherited responsibility for the third London airport, just four months later, dealt with the Government evidence with a sharpness which seemed at times to fall not very much short of contempt. The substance of the Report would, in any case, have made unpalatable reading for the Government. The local objections, he wrote, were 'formidable and justified'. Under the heading 'Opinion' he said: 'It would be a calamity for the neighbourhood if a major airport were to be based at Stansted. Such a decision could only be justified by national necessity. Necessity was not proved by evidence at this inquiry.'

The Report assessed the Government proposals on six counts and rejected them on five. There was no objection to it on air traffic grounds. But there were strong arguments against it on the grounds of town and country planning; bad ground access from London; noise; change of character of the neighbourhood; and the loss to agriculture. The criteria set out by the Inter-Departmental Committee had said about noise: 'the airport should be sited so as to disturb as few people as possible.' Clearly, this test had hardly been met by the Stansted proposal. The Ministry, Blake pointed out, had conceded, for instance, that of forty-three schools in Bishop's Stortford and its immediate neighbourhood half would be unable to continue if the airport were working to capacity, while others might be saved by remedial action. The Herts and Essex Hospital would have to close and 7,000 houses would suffer great nuisance. It was true, of course, that the noise could be mitigated by double glazing. But as one objector had protested at the inquiry: 'how can I soundproof my garden?' The Inspector could not say a great deal about the regional planning implications, since no detailed evidence had been submitted by the Government at all. But 'all the evidence submitted to me was that development of this kind in this area would be bad regional planning'.

Blake went on to deliver a series of short and simple lectures to the Government on how they ought to tackle a job of this kind. Indeed the whole tone of the Report seemed to carry the unwritten message: 'take this away, and do it again, and don't bring it back until it has been done properly.' No origin and destination survey had been made for passengers using Heathrow and Gatwick:

> Before deciding the location of the new airport it is, I think, essential to know where passengers are coming from and where they are going to. Insufficient attention has been given to the fact that the problem lies as much on the ground as in the air . . . [And again:] I suggest, when a passenger survey has been made, a review of the national pattern for the airports be made. If London is the inescapable focus, the location of the airport, the collection and distribution of passengers (which will require the co-operation of the airlines who own and run the terminals) should be considered with a target of preventing anyone who does not need to enter central London from doing so. [The Inspector concluded:] In my opinion, a review of the whole problem should be undertaken by a committee equally interested in traffic in the air, traffic on the ground, regional planning and national planning. The review should cover military as well as civil aviation.

Behind Blake came his colleague, the Technical Assessor, Brancker. He did not sound quite so cross; but he was still authoritatively critical. Like the Inspector, he clearly thought the whole job of choosing and assessing the potential airport would have to be done again: 'I am extremely hesitant to suggest anything which may lead to delay, but much of the evidence submitted seemed to me rather superficial and I would be very much happier to see a general examination in more depth before any firm decision is taken.'

Brancker did not even spare the evidence on air traffic control. In an assessment of the evidence which closely recalled the doubts expressed by the Inspector at Gatwick on air control evidence from the same source, he suggested that the Ministry might have been too conservative in ruling out some sectors around London from consideration altogether. It was quite clear, he said, that in approaching the whole problem, the Ministry of Aviation and the traffic-control experts had adopted a very conservative attitude and had insisted on a general separation of traffic streams, even in circumstances which might occur relatively infrequently; 'although in evidence however, Captain Hunt himself admitted that there would certainly be changes before a final pattern was adopted, it seemed to me that there was a tendency to treat the existing airways and sequencing areas as virtually immutable, whereas to obtain the best results a complete revision may be necessary'.

Elsewhere, Brancker marched firmly in condemnatory step with the Inspector. There had been inadequate information on the origin and destination of passengers. There had been inadequate information about alternative sites; Stansted was the only site which had been covered by a complete outline plan and even this did not include the costs. There was inadequate evidence, indeed almost no evidence at all, on the estimated growth of the air freight traffic – a form of traffic currently growing faster than passenger traffic.

The Inspector and the Assessor were also united in the deepest scepticism about the evidence of the Ministry of Defence, which, by insisting on the absolute need for maintaining the ranges at Shoeburyness, had effectively ruled out some of the alternative sites. The traditional restraint of civil servants in questioning contentions by the Ministry of Defence on grounds of national security did not bind Brancker and Blake: 'Evidence was given,' wrote Blake in his terse, commonsensical way, 'by Major-General Egerton that removal of the range could not be expected. I have, of course, to accept the evidence although the main thing about Shoeburyness is that it is there. It is difficult to think of a less suitable location for an artillery firing range than the Thames estuary.'

Brancker was not impressed either: 'The evidence presented for its retention was far from convincing and its presence on the doorstep of London appears to have been a complete anachronism.' But he took the argument a stage further, linking it with the possibility of an airport at Foulness.

> Little evidence was given about Foulness because it would be quite impractical to consider it unless the Shoeburyness range was removed. If this were done, the site would appear to have much the same characteristics as the proposed site at Sheppey. My personal opinion is that satisfactory air traffic control patterns could be devised and, like the noise problem in relation to Southend, it is capable of solution.[2]

The tone and conclusions of what Blake and Brancker had written faced the Government with a decidedly awkward problem. In the event, it was decided to take up the recommendation for a more wide ranging inquiry than that of the Inter-Departmental Committee; but this was done in a way which was far from the spirit of what Blake had seemed to be recommending. The inquiry would not be a public one; nor would it involve anyone outside Whitehall. Instead, the Government decided to appoint a further inter-departmental committee – although a wider-based and more high powered one than its predecessor.

This Committee was never named, and its Report was never published. The Chairman was Arthur Peterson, a Deputy-Secretary at the Department of Economic Affairs, and the other members were drawn from the Ministry of Housing and Local

Government, the Board of Trade, the Treasury, the Ministry of Defence, the Ministry of Transport, and the Ministry of Agriculture. The fact that none of them was identified – indeed the very existence of the Committee did not become known until long afterwards – naturally inspired the suspicion that the review committee might largely have been the first inter-departmental committee in disguise. This was later denied in the Lords debate on Stansted by Lord Beswick, the junior Aviation Minister of the early fifties who was now a Government Whip; the only two people who served on both inquiries, he said, were a technical witness and the secretary. But even Lord Beswick was unable to explain why the names could not be given. It was, he said, a tradition; though it was not, it appeared, one he was ready to go to the lengths of justifying. No one pointed out the strange circumstance that while the identity of the people on the second committee had to be kept secret, the names of the members of the first, civil servants as well as outside representatives, had been published at the beginning of their report.

But why should this have been a private inquiry, carried out by Whitehall, rather than a commission including outside experts? Had the terms of the Inspector's strictures, and the nature of his recommendations, been known at the time he delivered them, there might well have been demands for something more on the lines of the subsequent Roskill Commission. But the Government had decided not to let the Blake Report be seen until the second Whitehall committee had completed its re-examination.

This decision to make the inquiry a Whitehall matter was defended in the subsequent White Paper which was based on the findings of the Peterson Committee.

The Government [said paragraph forty of the White Paper] decided against an independent commission. In the first place, a decision either for or against Stansted was becoming urgent. Furthermore the Inter-Departmental Committee on the third London airport had already made a thorough examination of the question in 1962 and 1963; the subsequent publication of their Report and the public inquiry gave an ample and well-used opportunity to all interests outside the Government to put forward their own points of view. Accordingly, nothing useful seemed likely to be achieved by initiating a further round of public discussion of the same material. Not only would such a procedure inevitably have been very time-consuming, it would also have involved both the Government and all the other interested parties in renewed work and expense preparing and submitting evidence; and at the end of the day the decision would still have been for the Government to make.

It was not only those involved in the thick of the battle who found these assertions difficult to reconcile with the assurance which they thought they had wrung from Roy Jenkins that a further inquiry would not be denied them on the grounds of urgency.[3] What is more, the argument that such an inquiry would merely initiate a further round of discussion of 'the same material' rather seemed to imply that the Government would have had nothing to add at such an inquiry to the evidence presented at Chelmsford, which both the Inspector and Technical Assessor had agreed was thoroughly inadequate. In any case, naturally enough, there was quite a lot of new material incorporated in the result of the further inquiry to which the objectors would now have no chance to reply.

A more ingenuous, but probably more accurate, explanation was to be given to the House of Lords later by Lord Kennet, a Parliamentary Secretary at the Ministry of Housing. Lord Kennet did not attempt to pretend that the matter had been dealt with properly. 'I must admit,' he said, 'the way in which consultations and inquiries in this whole history have been handled has not been as good as could have been wished; but, and this is the point, it has been as good as can be achieved under existing law and practice.' His explanation for leaving the second inquiry to the people in Whitehall was simply that the people in Whitehall knew best: 'the Government are better placed than anybody else to take all factors into account.' There was a shout of 'Why?' at this point and Lord Kennet replied:

> Because the Government are at the centre and are able to call for facts from all quarters . . . Because the Government has more staff available, and because that staff has been working uninterruptedly on this problem for the last six years, we must accept that these figures reflect the true reality of social costs and social benefits more closely than any other figures we have got or shall get.

And with still more breathtaking candour Lord Beswick, adding his own explanation of why a public inquiry was impossible, said at the end of the debate: 'If the inquiry again came out in favour of Stansted, we should be all right as a civil air power; but if it did not, we should be sunk.' It was certainly difficult after that, to see that anything very much was left of the original Roy Jenkins pledge. And it was ironic that this argument of overwhelming urgency should come from the same source which, at the start of the decade, had been planning to postpone any decision on the third London airport for perhaps as much as five years.

Even had the inquiry by the review Committee been a model of thoroughness and accuracy and precision, it would still, because of the long history of distrust and grievance which had now grown up around the Stansted project, have been treated with suspicion both by the local objectors and by other critics of the Government's handling of its airport location policy. And the White Paper of May 1967 which embodied its findings did little to allay such suspicion by the way it went about its task. 'The Inspector submitted his Report by the end of 1966,' it observed blandly. 'He did not give a firm recommendation either for or against development of Stansted as London's third airport.' Anyone who turned from this to the Blake Report itself, which was released on the same day as the White Paper, could hardly have accepted that as a totally dependable summary of Blake's attitude to the evidence put before him at Chelmsford.

7

It should be said, however, that the report of the second Committee was an advance in sophistication on anything which the Government had achieved in this field before – and indeed, there was some pride in Whitehall about the level of expertise which it had brought to its work. The team had examined, along with Stansted, all the alternative sites which had been discussed before the Inspector at Chelmsford and had come up with two further possibilities of their own. One of these was Thurleigh, eight miles north of Bedford, where there was a Ministry of Technology airfield; the other was Silverstone.[4] The Committee proceeded to match all these other possibilities against Stansted, which they described as a 'bench mark', and to score their advantages and disadvantages. None of Stansted's competitors came out very well from this process. Apart from Southend, five Thames Estuary sites were examined. One of these – Gunfleet Sands, five miles off Clacton – had already been dismissed as unworkable in the Report of the Chelmsford inquiry. Others were Dengie Flats and Foulness, both in Essex, and Cliffe and Sheppey in Kent. Of these four, Sheppey was rated the most promising. The noise problem would be relatively small and the agricultural land taken would not be of high quality; the loss of rural amenity might not be so grave as with some other sites. But there were serious drawbacks. Access by road and rail was bad. The use of Southend Airport would be severely restricted. And the ranges at Shoeburyness would have to be closed and provided elsewhere.

How deeply the second Inter-Departmental Committee had pressed the Ministry of Defence on the value it attached to Shoeburyness it was impossible to say, since the White Paper did not explore the point and the Committee's own Report remained secret. But they had in the end, it seemed, come to accept what the Ministry was saying with the same complete trust which their predecessors had shown.

The facts are these [said the White Paper]: The Ministry of Defence will require the facilities at the range for as long as they can foresee; it is far from certain that the peculiarly suitable conditions at Shoeburyness for the work carried out at the range could be repeated anywhere else in the United Kingdom; lastly, if it were to prove possible to reprovide the facilities elsewhere this would cost approximately £25 million.

This figure of £25 million was in no way explained, and could for all anyone knew have been conjured out of thin air. Other costings which were also used to condemn the proposal to site the airport at Sheppey were equally extremely sketchy. The alleged cost of £40 million for providing rail access to Sheppey was, in particular, greeted with much disbelief; and it was further noted that, while a sum of £25 million for re-locating Shoeburyness had been included in the estimated cost of an airport at Sheppey, the cost of re-locating Wethersfield was not given in the White Paper at all. The approximate cost of developing Stansted with two parallel runways was said to be £47 million 'plus the possible cost of replacing facilities at Wethersfield'. The White Paper did not pretend that these were totally dependable cost figures. Such as they were, however, they did enable the Government to decide that 'for as long as it is possible to envisage better alternatives, Sheppey's grave defects in accessibility and high costs must rule it out as a site for a third London airport.'

The Committee's own discoveries, Silverstone and Thurleigh, did not finish the course either. Silverstone was rated better than Stansted in terms of regional planning, noise, and loss of agricultural land, but, as has been seen, it would have entailed shutting down eight RAF and USAF airfields and restricting activities at thirteen others. The cost of replacing them would be 'very substantial', perhaps as much as £100 million. The choice of Thurleigh would also have a 'detrimental' effect on military flying, and unlike Silverstone, it had no attractions in terms of regional planning.

The White Paper then embarked on a gentle demolition of specific points raised by the Inspector and his Assessor. There had been a misunderstanding of the evidence put to them on air traffic control; far from being 'conservative' in nature, it dealt with a pattern which was still in the experimental stage. It was said that the journey times to Stansted stated by the Inter-Departmental Committee and put before the inquiry by Government witnesses were suspect; the Inspector said that road traffic might well be delayed at peak times. But that was just as true of other sites as well. The Government was still convinced that of all the sites which appeared to be possibilities Stansted on this count was the one at which the difficulties 'would almost certainly be at their least'.

The Inspector had been wrong about the noise issue too. In fact, his presentation of Ministry of Aviation evidence was 'to some extent distorted and open to misinterpretation'. The White Paper pointed out that the noise nuisance at Stansted at peak times would only be about one-twentieth of what it would be at Heathrow – although it would be twice as bad at Stansted as at Silverstone and four times as bad as it would be at Sheppey. To the Stansted protestors, this comparison with Heathrow was like being told that a punch in the jaw was preferable to a stab in the back; especially when Sheppey, on this assessment, could have got away with what was relatively little more than a scratch.

And so, having reassessed the evidence, the second Inter-Departmental Committee concluded that the choice of the first Inter-Departmental Committee was right. Stansted was the best solution. Yet this was a rather different Stansted Airport from that envisaged in the previous report.

> The great difficulties experienced in providing and agreeing a site for a third London airport make it desirable [said the White Paper] that this third airport should ultimately be capable of development to the largest practicable capacity – namely to that of two pairs of parallel runways.

This was an eminently reasonable contention. One of the few lessons which was held to have emerged clearly in the history of airport development in Britain and in other countries was that to choose a site where further expansion was likely to be impossible was a clear recipe for trouble. Indeed, this very point had been emphasised by the Inspector at the Chelmsford inquiry. Setting out his own criteria for the choice of an airport, Blake had numbered among them the availability of enough surplus land to allow expansion. This would also, he said, keep new homes, new schools and other potential sufferers out of the area.

It was one thing, however, to accept the wisdom of this contention. It was altogether another to accept that a much bigger area of Essex might be invaded to make room for the new airport than had been envisaged at the time of the first Inter-Departmental Committee report, or than had been debated at the public inquiry. Yet there was, it was now claimed, no room left in the timetable for any further public debate. The White Paper was quite sharp about this.

It must now be clear that all the many and complex issues have been thoroughly scrutinised both within the Government and in public. Now that it has reviewed the various arguments for and against the Stansted proposal, the Government believes that the time has come when a decision must be taken and that the decision should be to go ahead with Stansted, as the best of all the alternatives.

The Government, therefore, proposed to obtain through Parliament 'public endorsement' of this conclusion. It would do so by bringing in a Special Development Order under Town and Country Planning Act of 1962 to confer the necessary planning permission on the British Airports Authority; this Order would shortly be laid before Parliament, and the Government intended to allow an opportunity for a debate on the Order and on this White Paper. But of any scope for the local and national objectors to contest the new evidence and new figures in the White Paper, or to discuss the fundamental change which had appeared in the concept of the airport as put before the Blake inquiry, there was no sign at all.

The day which the Government chose to produce its White Paper and to publish the Blake Report along with it – 12 May

1967 – was the day of the Whitsun adjournment of Parliament –
a day when, as is usual in these cases, many Members had already
left for their holiday. There were enough there, however, to
ensure a brisk skirmish with the Minister. Peter Kirk (Con.,
Saffron Walden)[5] recorded the 'deep resentment and bitterness'
which his part of the country would feel and Stan Newens
(Lab., Epping) talked of the 'depth of dismay' which would be
felt at the Government's 'stupidity and shortsightedness'. Study
the White Paper, Jay advised Newens; you may find when you
do that you'll change your mind. Eldon Griffiths (Con., Bury
St Edmunds) described the Government's decision as a classic
example of putting the convenience of aeroplanes before the well-
being of people and sacrificing the long-term interests of the
nation to a short-term palliative to the Airports Authority.[6]
Douglas Jay was having none of that.

> The fact is that if this nation is to hold the same place in
> civil aviation in the future which is highly important from the
> point of view of earning and saving foreign exchange, as
> we have held with shipping in the past, we must undoubtedly
> have as good a modern developed set of airports as our
> major competitors and that we are determined to have.

With which hurried summoning of the economic imperative,
the House passed on to other business.

The apparent determination of the White Paper that the objectors had been given a fair run, and should now shut up, only
increased the determination of the anti-airport groups to continue the fight. Five days after the announcement in the Commons,
the setting up was announced of a body known as the Stansted
Working Party. This resulted from a conference of MPs, representatives of local authorities, amenity organisations, the Preservation Association, The National Farmers' Union and the
Essex Branch of the Country Land Owners' Association. They
chose six people to be the spearhead of the continuing fight at
the political level against the airport proposal. They were Peter
Kirk and Stan Newens (the two MPs whose constituencies were

most centrally affected), Brigadier Collins of Essex County Council, Miss Lloyd-Taylor of Hertfordshire County Council, John Lukies (one of the Chairmen of the Preservation Association) and John Walker (representing the NFU). The group was staffed by a secretariat provided by the County Council. This group now set to work and produced one of the most formidable documents of the whole Stansted controversy. The setting up of the private Whitehall review had, they argued, been not only unfair and unsatisfactory but very nearly unconstitutional. Only the fact that the inquiry at Chelmsford had been on a discretionary rather than a statutory basis saved the Government from charges of serious constitutional malpractice. 'This was the most important planning inquiry of the decade,' said the Stansted Working Party. 'If it had been held in the normal way under section 22 or 23 of the Town and Country Planning Act, 1962 which it might well have been, the rules evolved as a result of the "chalk pit case" would have applied.'[7]

The town and country planning inquiries procedure rules of 1965 laid down that, if proposals were substantially changed after an inquiry had taken place, the parties must be told and must be given a chance to demand the reopening of the inquiry. This had in fact happened in two recent cases, one involving a hover-craft base at Pegwell Bay in Kent and the other, the North Sea gas terminal at Bacton in Norfolk. If this was thought necessary in these cases, why, asked the Working Party, had it not been done in the Stansted case too? The Stansted inquiry had proceeded on the basis of two runways on the airport with the distant prospect of a third. The White Paper envisaged two pairs of parallel runways; and this could only mean that the effects of noise on the surrounding area and the loss of agricultural land involved, must be greater than had been the case when the inquiry was convened. 'We feel,' said the working party, 'that the alteration is in effect to site both the third and the fourth London airport at Stansted.'

They also challenged the decision to set up the second Inter-Departmental Committee. It was as if the prosecution in a court case, having lost its battle before the lower court, had been given a chance to act as the court of appeal and get the verdict against it reversed.

> There is no assurance that the civil servants who constituted the Inter-Departmental Committee on the third London airport or who appeared as witnesses at the inquiry in support of the Stansted proposal did not play a major part in challenging the Inspector's review and in the work of the further review. Can it reasonably be expected that the Government departments which, through their representatives on the Inter-Departmental Committee, selected Stansted and defended the choice through the thirty-one-day-long Stansted inquiry, often by calling witnesses who had served on the Inter-Departmental Committee, can come to this review of the Stansted decision with an entirely unbiased approach?

The cost figures in the White Paper were described as seriously incomplete. What was needed was a cost benefit evaluation of all the selected alternatives. The failure to carry this out was, indeed, only one symptom of the hopelessly small scale way in which the Government had gone about its task.

> We have been concerned in establishing the feasibility of a number of major projects in many parts of the world [said the Working Party's consultants, Sir William Halcrow and Partners] and consider it useful to compare the Stansted proposals with the standards of preparation normally required by international spending authorities before irrevocable decisions are made. It is customary to require:
> (a) a broad assessment in economic terms of the place of the project in national and regional development and its viability within the global prospects in the industries on which its profitability depends;
> (b) cost and benefit studies for the project itself, and for a number of alternatives and variations;

(c) to substantiate (b) above, reasonably detailed schemes and layouts, with engineering estimates of construction costs, and comparative benefits discounted for an appropriate period of operation – generally at least fifteen to twenty years.

(d) a projection of capital expenditure phased with expected benefit and running costs in the recommended scheme.

'This is perhaps the biggest single planning decision which a Government of this country has ever had to make,' the Working Party concluded. 'It is proposing to site the major inter-continental airport of this country with all its geographical, social and economic repercussions simply on the most convenient World War II runway still extant. It is contended that this is no way to go about such a task.'

Already, before the Stansted Working Party Report, Essex County Council had decided to pursue the matter through the Courts. On 31 May they took out a writ against the Ministry of Housing and Local Government. In a High Court hearing from 24 to 26 July, the County Council asserted that the Minister was bound to act in accordance with the requirements of natural justice, and alleged that the intention to proceed by means of a Special Development Order, so avoiding a further public inquiry, did not conform with that duty. The Court accepted the Government's contention that this was a matter 'within the political arena' which it was for the Government to decide.

And now, a new and unexpected campaigner joined the ranks of those opposing the airport. The Technical Assessor at the Chelmsford inquiry, J. W. S. Brancker, wrote to *The Times* on 29 July 1967 to register his own objections to what the Government had now decided to do. He was then commissioned by the Preservation Association to produce a close analysis of the White Paper which appeared in October under the title of *The Stansted Black Book*. This demonstrated Brancker's attachment to the possibilities of an airport at Foulness more sharply than his contribution to the Report on the Chelmsford inquiry had been able to do. Foulness, he said, had been brushed aside in the White Paper because of its effects on Shoeburyness and on Southend Airport. But it was exactly in this sort of case where a full cost benefit study would be valuable. Although Southend was a busy airport, contributing much runway capacity and making a profit, trouble had already arisen about the noise of jet aircraft. In the long term, Southend might be better off if the present airport were shut and traffic transferred to a fully developed airport at Foulness.

He went on to argue that the inquiry which had recently been set up into the future of aviation under the chairmanship of Professor Sir Ronald Edwards should be linked with an attempt to develop a national airport plan 'because otherwise it would be tantamount to planning railways without bothering about the stations – or even junctions and marshalling yards'.

All this, combined with reports from the beleaguered area of protest meetings, emergency fighting funds, car stickers being printed and the rest, received wide and sympathetic publicity. But, it appeared, as will be seen, to be having very little effect on the Government, and especially on the principal defender of the Stansted decision, Douglas Jay. The Government believed – and in theory no doubt there was considerable justification for this view – that the resources it had brought to bear on the question of the location of the third London airport outgunned anything that the critics might produce. After all, however formidable the objections raised to Stansted might have been, the objectors had done themselves little good by producing as an alternative a site at Padworth whose claims evaporated on any close consideration.

But this was not really the point. The long series of botched or badly presented decisions and of cases inadequately made, the secrecy, the sneaky way in which the Blake Report had been held back, and the bombshell dropped on the eve of a holiday in the form of the White Paper – all this meant that anything the Government did by now was likely to be deeply suspected. To the natural human sympathy which was created by reports of

ordinary people battling in defence of their threatened homesteads was added a measure of respectful attention for the quality
of the arguments they were producing and for the apparent
imagination of their latest alternative solution. In the contest for
the support of people who were initially uncommitted on the
Stansted issue, the protestors now appeared to be building up a
steady and probably invincible lead over the Board of Trade.
Organisations like the Town and Country Planning Association
and even the South-Eastern Economic Planning Council – which
the Government had failed to consult about its decision to stand
by Stansted – were now lining up alongside the protestors. After
a Shadow Cabinet meeting on 15 June the Conservative Leader,
Edward Heath, called for a halt to the Stansted project until a
full public inquiry had taken place.

That was only to be expected. The Conservatives had been
saddled with the blame for the inadequacy of the first Inter-
Departmental Committee inquiry; it was natural that they should
seek to restore some credit now by challenging the unsatisfactory
nature of Government policy since Labour had come to power.
What was much more important to the Government, however,
as it assessed its chances for getting the Special Development
Order through Parliament, was how its own troops would
behave when the vote came to be taken in the Commons.
There were two occasions in the 1966–70 Labour Government
when back-benchers successfully deprived the Government of
legislation on which it had set its heart, and it is instructive to
compare these two revolts which succeeded with the much publicised Stansted revolt which never in fact took place.

The two real-life revolts were on the industrial relations legislation, which followed Barbara Castle's White Paper *In Place
of Strife*, and on reform of the House of Lords. In the case of the
industrial relations legislation, the hand of the rebellious back-
benchers was strengthened by two crucial circumstances; the
unmistakeable hostility of large sections of the Party and of the
trade union movement – the Labour Party's National Executive
Committee took the unusual step of repudiating *In Place of*

Strife – and the knowledge that the Cabinet itself was deeply divided, with one of its senior members, James Callaghan, having strenuously signalled his wish to see the legislation abandoned. A militant minority within the Cabinet; a revolt by the trade unions; entrenched opposition among the back-benchers – any of these alone might have been within the power of the Government to defeat, but all these circumstances together produced the memorable scene, graphically described by Peter Jenkins in *The Battle of Downing Street* (C. Knight, 1970), when the Chief Whip, Bob Mellish, told the Prime Minister that there was no longer any hope of getting the measure through the House.

The other revolt, on Lords reform, had, as Jenkins shows, turned out in a way to be a dress rehearsal for the much more serious attempt to sink the Industrial Relations Bill. In this case, back-bench members of both Parties, among them such eminent figures as Michael Foot and Enoch Powell, united to defeat a measure which, as they saw it, was being forced on the Commons by the collusion of the front benches.

In each case, the significant circumstance was that a large enough number of the Government's back-benchers felt strongly enough against the measure to go further than merely signalling its distress and to show itself ready actually to defeat the legislaton. It seemed for a time in 1967 – two years before the struggles over *In Place of Strife* and Lords reform – as though a similar revolt would occur over Stansted. On 14 June Peter Kirk and Stan Newens, with four other MPs – the former Liberal leader, Jo Grimond; the Conservative Member for East Herts Sir Derek Walker-Smith, who also had a close constituency interest; Tom Driberg, Labour MP for Barking and a member of the National Executive; and Selwyn Lloyd, former Foreign Secretary and subsequently Speaker of the House – put down an Early Day Motion demanding that before any irrevocable decisions were taken on the siting of airports, a new and realistic estimate of future needs should be prepared, with full consideration not only of the probable development of air traffic, but also of good communications, agriculture and amenities of the country-side. It asserted that, given proper use of existing airports, there was time enough to prepare such an estimate; and it called on the Government not to proceed with the major expansion of Stansted until the matter had been fully investigated by a Royal Commission or some other appropriate body.

The Early Day Motion is one of those parliamentary devices whose true function bears little resemblance to its name. One might assume that the purpose of MPs in putting Early Day Motions is to get them debated on some early day. But, in fact, that hardly ever happens. Most of these motions simply lie on the Order Paper, gradually accumulating names. If a motion is tabled in the names of the Opposition front bench, then time will probably be provided for it. But if it is simply the work of a collection of back-bench MPs, however eminent or numerous they may be, no debate will normally take place.

Many of these motions are simply a way of letting off steam. Some of them are simply exuberant congratulations to successful football teams or celebrations of some other constituency triumph. The more serious ones provide MPs with a useful way of demonstrating their strong feelings on a given issue. In particular, they are a valuable device for allowing back-benchers on the Government's side to indicate to their leaders that they do not very much like what the Government is presently up to.

The Stansted Motion managed to accumulate no fewer than 234 signatures on the day it was put down. This is even a more formidable total than it seems. At least a hundred of the 630 MPs in the House do not normally sign these motions because they are members of the Government, while the senior members of the Opposition front bench usually keep out except on special occasions. So the headlines next day which described the Early Day Motion as an explosion of anger were quite understandable.

For the Government Whips, of course, what mattered was not the total number of signatures, but the number which had been put in by their own supporters. On the face of it, they seemed to be in some trouble. In all, 258 MPs finally signed the Early

Day Motion – one Liberal and one Conservative, in their enthusiasm, managed to sign it twice. Of these, ninety-one were from
the Labour side – a total which meant that if they all backed their
opinions with their votes on the Order, the Government's
majority of ninety-three would be comfortably obliterated and
Stansted would be safe. In fact, however, the Government paid
no more attention to the threat of a parliamentary demonstration
of no confidence in its decision than it had to any of the others.
There was indeed a clear division of opinion in the Cabinet about
the rightness of the policy, with Roy Jenkins and Anthony Crosland the two principal dissenters. But there was little prospect of
the policy being abandoned. There seemed to be overwhelming
reasons for continuing to steam on, whatever danger signals
appeared to be flying.

To begin with, the Cabinet was satisfied that the opinions of
Whitehall officials who had formed the second Inter-Departmental
Committee were more than a match for any of the outside
experts who had testified against Stansted. Ministers continued to reiterate to troubled back-benchers that the White
Paper had established an unarguable case for the policy it advocated. Just as Douglas Jay had, with sublime confidence, told
Newens that a reading of the White Paper might assuage his
fears about the proposals, so Harold Wilson also advocated a
reading of the White Paper to a man who wrote to him criticis-
ing the policy.

> The Prime Minister has asked me to assure you [said the
> reply] that this decision was taken only after a very thorough
> examination of Stansted itself and the various possible alter-
> natives. He feels sure that if you study carefully the White
> Paper which the Government has published, you will see
> that there are far stronger objections to each of the possible
> alternative sites than to Stansted.

Douglas Jay put it even more strongly, asserting, as he was
later to do in the Commons debate, that the talk of finding a
better site on the coast was so much wishful thinking; that the

White Paper had made a case with which no reasonable person could honestly disagree; and that the Stansted decision must now be regarded as irrevocable. An equally uninhibited assertion of faith in the Whitehall calculations came later in a Lords debate from Lord Kennet, who, having declared his dissatisfaction with the way the whole process of the Stansted decision had been handled, and having promised that things would work better in future as a result of the new Town and Country Planning Bill, went on to declare:

> I should not like to leave this subject without recording my own conviction – and I believe the conviction of all those who have been able to observe this matter developing at the centre – that if such a procedure had been in existence when the third London airport question was being considered it would not have come up with any other choice than Stansted.

After all, Ministers told themselves, the Ministry's experts had now spent nearly six years of continuous work on the choice of the third London airport site. No other experts, however, dedicated, could match that. What was more – unlike the people who appeared to testify at planning inquiries – the Ministry's experts spoke with the extra authority of people who would have to implement the decision once it had been taken; and who, therefore, bore all the more responsibility for getting their calculations right.

There was too, every reason why Ministers should be more aware of the urgency of getting the development of the third airport moving than their critics at Westminster and in the country. Efficient airports were essential to our trading position; and with the economic crisis building up which was to lead, only a few months later, to devaluation, that kind of consideration had rarely seemed more important than it did now. It emerged at the top of the Government's list of priorities in Jay's speech in the subsequent Commons debate. 'The main object of our airport policy,' he declared then, 'is to ensure that Britain retains its present European lead in civil aviation and that we remain ahead of our competitors.' Unless we held a substantial share of the growing world air traffic, our balance of payments was bound to suffer. And in order to hold this traffic, we must be able to offer to the public and to the airlines the first-class modern airports which they expected. 'A major effort,' Jay warned, 'is now being made by our main rivals to wrest from us the leadership we now hold in Europe and the tourist traffic and trade which go with it.'

The argument that the urgency was now so great that no further inquiry could be contemplated – the very argument which Jenkins had indicated would not be used – was now in full cry. It had been assiduously fed throughout the summer by Peter Masefield and his colleagues of the British Airports Authority. Masefield had met MPs at the Commons and warned them that already traffic was having to be turned away from Heathrow. In a letter to *The Times* on 28 April, just before the White Paper appeared, he had launched a major assault on Sheppey – which was to be identified by the White Paper as the best of the coastal sites. On 12 June – two and a half weeks before the Commons were due to debate the White Paper – Masefield held a big and widely reported press conference at which he returned to the theme of the absolute necessity to get started on Stansted straight away. It was, he declared, the only suitable site. It could outstrip Heathrow and could certainly rate above Gatwick as a major airport. 'None of us,' said Masefield, 'would like to have an airport stuck down in our back garden' – a remark which was not quite so complacent as some took it to be, for Masefield himself lived close to Gatwick Airport, at Reigate. But he did not accept that an airport must necessarily be a 'concrete waste'; Stansted could be made attractive by landscaping.

On 24 June, the BAA's Director of Engineering, Norman Payne, described much of the criticism of the Stansted decision as 'emotional, ill-informed and exaggerated'. 'We have more experience in these matters than anyone else,' he added. 'I have been building airports for fifteen years and I am confident that Stansted is the only realistic place to put a third airport for

London.' It would certainly take ten years to build an airport at Foulness – and it would be quite impossible to wait as long as that. On 17 July the Chief Executive of BAA – George Hole, who had chaired the first Inter-Departmental Committee – said that it would be 'practicably impossible' to find a realistic alternative to Stansted.

This barrage of exhortation to set sentiment aside in the face of national necessity died down a little after the Commons debate but it was in full swing again in time for the House of Lords debate on 11 December. In the flurry of correspondence which followed that debate, Masefield stated his claim in even bolder terms. In a letter to the *Observer* published on Christmas Eve, he wrote:

> We who have the task of providing this capacity have no less regard for the English countryside than any other Englishman. But – as each year shows more clearly – we shall survive economically as a nation only if we get our priorities right.
>
> A relatively remote estuarial airport, compared with Stansted, would rapidly go bankrupt and stay bankrupt. Either we must accept that we must run many aspects of the nation as a business in a competitive world – in which our communications and their costs are fundamental – and go all out to develop our trade and pay our way, or we choose to stagnate. The use of a small area of Essex to earn many millions of pounds of foreign currency a year – and at the same time pay its way – will do more to preserve our British way of life and the future of our country as a whole, than can any other comparable project.

Next in line was the *Financial Times*, which on 27 December printed a letter from Masefield declaring that Foulness would, on conservative estimates, cost at least £150 million to build and would cost £20 million a year more than Stansted to operate: 'The sacrifice of a relatively small area of sparsely populated land close to London has to be weighed with the priorities of our national economic life and communications as we move into the twenty-first century.' On 8 February as rumours continued to spread that the new President of the Board of Trade, Anthony Crosland, was contemplating reopening the whole affair, Masefield said: 'Unless work begins on London's third airport not later than the end of this year, Britain will be caught with her national and aeronautical pants down by 1974.' By 1974 aircraft movements would be nearly fifty per cent higher, passengers would have more than doubled and air cargo would have increased by 130 per cent. By the end of 1973, Heathrow would be filled to capacity. On 10 February, in a letter to the *Daily Telegraph*, he accused Sir Roger Hawkey, Co-Chairman of the North-West Essex and East Hertfordshire Preservation Association of 'a disservice to the country' in saying there was still time to spare for a further inquiry.

In the whole of this campaign, two spectres were repeatedly fetched down from the cupboard and paraded for inspection. One was the accumulating delay and congestion affecting New York – a potent warning of what might happen if necessary decisions were not taken in time. The other was the serene and impressive progress of the French towards their new airport at Roissy-en-France – for which design studies had begun as long ago as 1957. Here was the prospect of a fine modern airport, only some fourteen miles from the centre of the capital, which expected by 1975 to handle twenty-five million passengers and more than two million aircraft movements a year. If Britain failed to produce its own airport in time, the clear prospect was that France would take the cream of its market.

On both these scores therefore – the dependability of the existing decision, whatever outsiders might say about it, and the urgency of getting started – the overwhelming feeling among senior ministers was to let the Stansted decision stand. But it was, in any case, not a subject to which they could go on devoting valuable time.

Even for a government which lived on crises, these late spring and summer months of 1967, when the Stansted decision was first taken and then defended against every-growing opposition, were unusually fraught. It would be a hard man who did not sympathise as he read Harold Wilson's own account of trouble piling upon trouble as the year advanced. 'Those midsummer weeks of 1967,' he writes in *The Labour Government, 1964–70: a Personal Record*,[8] 'were without respite at home or more particularly abroad.'

At the start of the year, the Government was heavily preoccupied with Europe. Wilson himself and George Brown made their tour of Common Market capitals in January. From the end of March to the beginning of May, the Cabinet was meeting twice a week to thrash out every issue which the Market raised – and with such deep-dyed doubters as Douglas Jay in the Cabinet, there were many. 'If anyone had asked for a document on the effect of entry into the Market on British pigeon-fancying,' wrote Wilson afterwards, 'he would have got it; but no one did.' The draft statement on the decision to apply for membership went through the Cabinet on 2 May; at the end of a three-day debate on 8 to 10 May (the Stansted announcement was to be made on 12 May) there were thirty-five Labour abstentions. De Gaulle's press conference, apparently writing 'fin' to the whole enterprise, was on 16 May.

But a second epic struggle was now developing; the build-up to the Six Day War in the Middle East, which was to become 'the biggest contributing factor' to the subsequent devaluation. In mid-June there was a crucial vote on a new stage of the prices and incomes legislation, with forty Labour Members abstaining; and by now there were accumulating signs that the Government were heading for defeats on the policy at the TUC Congress and the Party Conference.

Later, there were the dock strike, the loss of Hamilton to the Nationalists in a by-election (the crowning moment in a cataclysmic run of local election and by-election defeats), the devaluation itself on November 18 and the development of a schism in the Cabinet and the Parliamentary Party, over arms to

South Africa, which Wilson describes in his book as more serious than any other such dispute in six years of Labour government.

Add to that such sudden and totally unpredictable crises as the D-notice affair, which began in February and blew up again with the publication of the Radcliffe Report in June; the Torrey Canyon, hitting the Sevenstones Reef in mid-March and setting up a problem of oil pollution the like of which no one in government had ever encountered; and the persistent demands of Aden, Nigeria, Gibraltar and Rhodesia on senior ministers' time; and it is easy to see why the third airport, even though it was, in terms of investment and planning, a decision of no mean importance, would tend to be relegated to the sidelines. It gets three lines in Harold Wilson's book: 'a grumbling row over the siting of London's third airport erupted into a full-scale campaign by the "not at Stansted" lobby; the issue had not been well-handled by the departments.' That sentence faithfully reflects the place of Stansted in the government's impressive collection of apparently insurmountable problems.

The Cabinet found time for a debate before Jay's announcement was made – according to one participant it ran for about an hour with opinion dividing roughly two to one in favour of soldiering on. On 15 June, a fortnight before the Commons debate and in the face of mounting hostility, the Cabinet looked at it again but again decided to stand firm. 'Some senior Ministers believe,' said *The Times* next day, 'that the whole principle of planning is involved here and that the Cabinet must demonstrate that there can be no planning on the basis of sentiment or retreat before every local and regional interest that may be involved.' In quieter times the sceptics might have fought longer and harder; but not now.

The Government's determination not to abandon Stansted was also fortified by the belief held by several ministers, and particularly by Jay himself, that the campaign against it was an overwhelmingly middle-class (and probably an overwhelmingly Conservative) affair. It really would, in any case, be wrong for the Government to go back on what it believed to be the right course in the national interest, simply because the pressure group opposing it happened to be eloquent as well as vociferous, and rich enough to mobilise enough money to maintain a campaign on a more than parochial scale. Some local people, Jay told the Commons, had wanted the airport so badly they were actually petitioning for it; and the Harlow Trades Council was taking a quite different view from the Hertfordshire and Essex County Councils.

This was an argument to which Labour members might be expected to show a particular susceptibility. And indeed, at the very moment when Jay was under fire in the country for being so inflexible, he was being criticised by some in the Labour Party for giving too much away. To some in the Labour Party – among them, Roy Mason, the robust miner's MP from Barnsley who was a Minister of State at the Board of Trade until January 1967 – the Stansted pressure group was a species which only existed south of Potters Bar and ought to be treated with the greatest distrust. These middle-class people with time to spare and money at their finger tips, some Labour members complained, were in a strong position to promote their own private interests at the expense of the poorer sections of their local communities and of the national interest as a whole. The Government should not be pushed around; the duty of governments was to govern. While some thought Jay too deaf to argument, others thought him too ready to listen. 'Every time I went into his office,' said one later, 'it seemed to be full of these major-generals and retired bank managers from Essex.' Jay himself spoke bitterly of the Stansted lobby in a debate in February 1973:

The public relations campaign that we had in 1967 against Stansted was not a spontaneous expression of local opinion. It was organised largely by a group of substantial landowners one of whom was simultaneously trying to sell some of the land in question to the British Airports Authority. The group spent £40,000 on a public inquiry and a further large sum in propaganda after it ... of the letters and postcards reaching the Board of Trade at that time, those against Stansted came mainly from W1 and SW3 addresses.

And the Government could produce another good socialist defence of Stansted. An airport at Sheppey would cost £31 millions more than Stansted;[9] an airport at Foulness, perhaps £35 millions more; one at Silverstone, as much as £62 millions more. How many new homes, how many new hospitals, ran the repeated rhetorical question, could be built for the benefit of ordinary people across Britain by adopting the cheapest solution, whatever squirearchal squawkings it might have set off in Essex and Hertfordshire?

It was arguments like these with which Jay and the Housing Minister, Anthony Greenwood, had held the Cabinet firm, and which were now offered as balm to the troubled consciences of back-benchers. And it seemed that standing firm would be enough to do the trick. Government business managers could see that the number of Labour MPs with any deep motivation to oppose Stansted was nowhere near big enough to give them serious trouble. Stan Newens, in Epping was clearly a deeply committed opponent. Renee Short, a former Hertfordshire County Councillor, who now represented Wolverhampton North-East, was also completely opposed. But Essex is not, even in the best of Labour years, good territory for the Party at elections. Of eleven seats within fifteen miles of the proposed new airport, there were only three constituencies represented by Labour MPs, and one of these – Shirley Williams (Hitchin) was a junior minister. The rest of the Labour Members who had signed the Early Day Motion might feel strongly that Stansted was wrong; but they did not have the additional stimulus of angry constituents back home who might fail to vote, or fail to work, in the next election if their member voted meekly in the Government lobbies on the Stansted Order.

The meeting of the Parliamentary Party on 22 June confirmed the accuracy of this assessment.[10] Fewer than fifty Labour MPs were present. Both Jay and Greenwood took a tough line;

they would not, they insisted, be moved. That was not greeted with enthusiasm; but there was certainly not the avalanche of protest either which would have presaged any serious revolt.

So by the time the Commons finally came to debate the choice of Stansted on 29 June, the explosion of anger which had once been forecast on the Labour side turned out to have little more political impact than the firing of a pop gun. Robert Carr, Chairman of the 1960–61 Estimates Committee and now Opposition spokesman on Aviation, moved that an independent committee of inquiry should be set up into a national airports policy, in the context of which a decision on the third London airport could be taken. This, he declared, was not to be read either as support for the choice of Stansted or criticism of it. As far as the Opposition were concerned, the verdict must be one of 'not proven'. Yet clearly, the decision-taking machinery had failed to give proper weight to the social cost, the national planning implications and the transport problems involved in locating a major airport. The way that the Stansted decision had been taken, Carr argued, had come to assume an importance greater than the decision itself; it was, in a sense, a touchstone for the future large scale planning decisions with which Britain was bound to be faced in a technological world.

The first Inter-Departmental Committee (a Conservative creation, it will be recalled) with its heavy aviation bias, said Carr, had been logical enough; it was no good, after all, planning airports you could not fly from. The real villain of the piece, in his view, was clearly the review Committee (a Labour creation); the White Paper based on their work had been 'an insult to intelligent people'. Why had the White Paper been satisfied with a weight of traffic movements on the runways of London's airports which was far below that sustained in the United States? Why couldn't the Government reach a clear decision, even now, on the future of Gatwick? Why had the costings been so vague? Why was the decision unrelated to a national communications plan? And why had there not been a full systems analysis? Clearly, Conservative aviation thinking on what was involved in planning an airport was more sophisticated now than when the first Inter-Departmental Committee had been set to work. 'We want,' concluded Carr, 'indeed we must have, the economic benefits and the personal opportunities [of flying] but in obtaining these benefits we must not become slaves to technology.'

But to Douglas Jay, our economic future depended on starting now; and that meant starting on Stansted. It also, Jay added, meant getting to work on Gatwick's second runway. It would, of course, be possible to examine ten other sites, but that would mean a full inquiry into each, lasting a year, perhaps more. By the end of all, we would probably be left with no airport, and no decision.

Jay then set out the comprehensive nature of the review Committee's operation and rehearsed the logic of its conclusions, though up-dating some of its figures; the Stansted estimate was now £55 million (the cost of re-siting Wethersfield having been belatedly added in) while Sheppey had come down a bit to between £120 and £130 million. Both governments, Jay declared, had now surveyed the territory and settled on the same site; so it must – to adopt Julian Amery's words four years earlier – be 'the right choice'. And surely, having spent six years on these inquiries we could not continue with more – especially if they were by outsiders. 'After all the investigations which have already been made,' Jay concluded, 'there is no more case for handing this decision over to an outside body than there is on the Government's decision on the level of pensions or the level of income tax, or for that matter the level of defence expenditure.'

Among Labour back-benchers, Stan Newens fiercely condemned the handling of the whole affair by the party which had come to power as a party dedicated to planning, and Renee Short also announced her intention to abstain. But Terry Boston, MP for Faversham in Kent, was equally militant against any threat to the Kent coast; and the seventy-seven-year-old MP for Glasgow Govan, John Rankin, looking back across more than twenty years of aviation debates, expected that the Stansted agitation would die away just as opposition to Heathrow and

Gatwick had done. And Robert Howarth (Bolton, East) who had been alarmed by Masefield's forecast of a saturated Heathrow, spoke strongly in favour of the Stansted decision.

So Anthony Greenwood, winding up, had no great number of potential rebels to convince on his own benches in order to ensure the Government's success. He began by an admission; the long assumption that Stansted would be chosen had come to dictate its eventual choice.[11] The Government were even now at work on better procedures for dealing with these great planning issues. The new White Paper on Planning Policy had been published that very day.

> Where issues of this sort arise, the ordinary public local inquiry is not satisfactory either as a method of permitting the full issues to be thrashed out or as a basis for a decision which can take into account the whole range of practical alternatives. The Government are, therefore, examining whether and how procedures can be changed for these exceptional cases while avoiding unreasonable delay or impairing rights of making objection and being heard.

Not one Labour MP voted against the Government in the two divisions which followed – the first on the Conservative amendment, which was defeated by 303 votes to 239, and the second on a motion approving the Government's decision, which was carried by 303 to 238. One MP – Michael Foot – who voted with Labour on the first division abstained on the second. Reports of the number of abstainers varied from three to ten. It is never easy to be sure about abstentions, since deliberate abstentions are not recorded, nor does Hansard contain any list of those who are absent but 'paired' with absentees on the other side. Frequently those who are asserted in the Press to have been deliberately absent turn up later, full of indignation at having been singled out when their loyalty was never in question. All that can be said with certainty is that of the ninety-one Labour members who had signed the Early Day Motion, seventy-four voted twice with the Government on the night, one voted on the first motion but not the second, and the rest did not vote in either.[12]

It was a performance which drew down, from the Conservative benches and from the protest movement outside, much obloquy on those Labour MPs who had not backed their signatures with their votes – or even with their abstentions. But the circumstances in which MPs with no constituency motivation could have been expected to engineer a really serious rebellion did not exist.

It was recognised that a defeat on Stansted would be a serious matter for the Government, adding one more embarrassing difficulty to the imposing pile already in its in-tray. MPs normally wish to reserve the weapon of abstention or a contrary vote either for an occasion when deep political issues are involved or for issues which are in some sense non-political or supra-political. If the airport were to be built at Stansted, there might be some well-merited derision from posterity; but it was not – as were the highly resented measures on prices and incomes, on prescription charges and school meals, and above all on the laws affecting the trade unions, which were to come later in that Parliament – an issue in which the very nature of the Labour Party, of socialism in Britain, could be held to be at stake. Nor, of course, in any strict sense, was Lords reform. But that, in a way, was less of a straight political issue than Stansted. The Opposition had a free vote on that occasion, so you did not have to risk the ill-fame of voting in their lobby, or to sit by and watch them jubilant at a Government defeat to which you yourself had contributed. Also, this was a parliamentary matter; and MPs recognise a loyalty to parliamentary institutions which cuts across their normal loyalty to Party.

So Douglas Jay's defence of the Stansted policy had paid off. But he was not, as it happened, to continue at the Board of Trade for long. His unconcealed and unrepentant hostility to British participation in the European communities had caught up with him. On 29 August – following a meeting with Harold Wilson on the railway station at Plymouth[13]; – Jay retired to the back-benches, making way, by one of those pleasing symmetries which attend the construction of an administration, for the man who had taken Education when Jenkins, because of his commitment to Aviation, turned it down: Anthony Crosland, the Labour MP for Grimsby who that day celebrated his forty-ninth birthday. From that moment, the course for which so many Labour MPs had so loyally voted at the expense of their personal predilections was to be given the rigorous challenge it required, and was finally to be abandoned.

7

. . . In which the Government bows to an unrepresentative elite, and decides to pass the problem to someone else

Anthony Crosland was an economist by training, and when Labour were elected in 1964, he had appropriately become Economic Secretary to the Treasury. So he was a natural enough candidate to succeed Jay – another economist – as President of the Board of Trade (the post which Harold Wilson had himself occupied under Attlee when only thirty-one). But Crosland was also a politician for whom the environmental issues had been a compelling interest long before they became a fashion. In *The Future of Socialism*, published in 1956,[1] he had boldly asserted this principle: that the look of our cities and countryside; the preservation of the best buildings of the past and the design of buildings which would decorate rather than demean the townscape; the cultivation of gaiety and style in our national life rather than Fabian asceticism, were ideals which politicians ought to be pursuing along with the traditional crusade to obliterate poverty and build a just society.

This theme of his book was not the least suspect thing about it. Its revisionism – it became the bible of the Gaitskellite Social Democratic young – would anyway have caused deep doubt and resentment about him on the traditional Left. But the assertion of this wider view of what politics was all about caused some to question his very seriousness as a politician. It was often said, even by those who extolled his grasp and his practical capacity, that he lacked that power to drive himself, to shut out the pleasures of the outside world, which some regard as the essential mark of a truly serious politician.

He was, in time, to be further promoted. From October 1969 until the fall of the Labour Government, Crosland was Secretary of State in a new department of Local Government and Regional Planning, a sort of umbrella ministry which enfolded the main environmental departments of housing and transport, kept a supervisory eye on the regional planning boards and councils, and dealt with local authorities on the tender question of local government reorganisation.

This was the time when the politics of the environment were beginning to take an increasing hold on the imagination of the nation – though more obviously so in the affluent, restless world of middle-class London suburbia than (to use Crosland's own endlessly reiterated example) among the back streets of his own Grimsby constituency. Crosland, said the more militant environmentalists, was not serious in his commitment to their cause; he was riddled with compromise; he did only what could be achieved without political discomfort. He showed no necessary urgency about population growth, he continued to acquiesce in the possible loss of areas of superb natural beauty to the rapacious demands of industry. Even Peter Walker, the epitome of self-made business success, who became environmental overlord in the Conservative Government, was more to be trusted, it was said, than Crosland, who now sat across the despatch box from him as Shadow spokesman on the environment.

It was an allegation which Crosland sometimes dismissed wearily, sometimes snapped at, out of a mixture, it seemed, of boredom and anger. In a Fabian pamphlet, *A Social Democratic Britain* (1971), he hit back at those who had been condemning him on these grounds. He still rated an increase in economic growth at the very top of his list of political priorities. (It was one of the ironies of the discussion about the sacrifice of environmental values to economic growth that it broke out at a time when the rate of economic growth in this country was negligible anyway.) Without it, he said, the necessary objectives of socialists in Britain, the creation of a more generous and more just system of welfare, would prove to be impossible.

I do not of course mean that rapid growth will automatically produce a transfer of resources of the kind we want; whether it does or not will depend on the social and political values of the country concerned. But I do assert dogmatically that in a democracy low or zero growth wholly excludes the possibility. For any substantial transfer then involves not merely a relative but an *absolute* decline in the real incomes of the better-off half of the population (which incidentally includes large numbers of working class voters); and this they will frustrate.

And turning on the extreme environmental crusaders, he declared:

Their approach is hostile to growth in principle and indifferent to the needs of ordinary people. It has a manifest class bias, and reflects a set of middle- and upper-class value judgments. Its champions are often kindly and dedicated people. But they are affluent and fundamentally, though, of course, not consciously, they want to kick the ladder down behind them. They are highly selective in their concern, being militant mainly about threats to rural peace and wildlife and well-loved beauty spots; they are little concerned with the far more desperate problem of the urban environment in which eighty per cent of our fellow citizens live.

Crosland's own doubts about the Stansted decision – he was well acquainted with it, having entered the Cabinet when he became Minister of Education in 1965 – were not, as some have assumed, simply the result of a more sensitive concern for environmental values than those demonstrated by his Cabinet colleagues. It was not, as his Fabian pamphlet indicates, that he valued economic growth any less than they did. What equally concerned him was the failure, in reaching the Stansted decision, to make that kind of calculation which any economist would have seen was necessary. It was not just that an area of Essex stood to be blighted; it was also that it might be blighted because of a decision which had never been properly assessed, or treated to the processes of economic analysis which were now becoming common place. Crosland determined that – as far as the resources of his new Department permitted, and they were sadly limited – such a re-examination would now take place. Later in conversation with Maurice Kogan, published in *The Politics of Education*[2] he described the department's reaction:

> One of the few times I ever had bad relations [with civil servants] was when I first went to the Board of Trade and said I proposed to reopen the Stansted decision. I shall never forget the row of black faces! The officials thought they had finally got Stansted, and were enraged to have the whole wretched matter reopened. They thought I was mad and wrong. My relations with that group of people were pretty bad for some time. But that was an unusual case because it was reopening a decision which it had taken them years of work to arrive at.

Ironically enough the step which most immediately persuaded Crosland to reopen the case had been authorised by Douglas Jay. The Inter-Departmental Committee's failure to apply any kind of cost benefit analysis – a course which had been urged by many of the Government's critics – had been noted with dismay by the head of the Department's Economic Services Division, John Heath. Heath himself was, in a way, a new kind of man to find in the Board of Trade; three years before, he had become virtually the first full-time house economist to serve in the Department, when his unrelated namesake, Edward Heath, then President of the Board, had accepted the argument that it was high time that the Board should have an economic unit of its own. (That a department so centrally concerned with the economic life of the nation should, as late as 1964, have been operating without this kind of expertise was one of the circumstances which had led Harold Wilson in his opposition days to condemn the Civil Service addiction to classical scholarship rather than scientific skills, the tendency to allow the gentlemen rather than the players to conduct our national affairs.)

John Heath now suggested that even at this stage the calculations would be worth doing. Jay, though reaffirming his faith in the rightness of Stansted, agreed. Working with a staff of two, Heath examined some half a dozen sites, some of them unscrutinised in any of the previous trawls for a London airport site. The best of the lot, on the calculations of Heath and his team, appeared to be near Cranfield in Bedfordshire. That didn't enable him to recommend that Cranfield should be chosen; obviously, studies in much greater depth and detail would have been needed before it was possible to make as firm a contention as that. What it did clearly show, however, was that the argument that Stansted was the only realistic choice didn't stand up to close examination.

Then Jay left; and the results of Heath's researches came up before Crosland. He immediately decided that the new evidence was important, and justified more work being done. So other departments were called in for consultation. The trawl of sites was widened: Foulness, which had been disregarded in the early stages because of the Ministry of Defence objection, was brought into the field; so was another site to be listed later by Roskill – Thurleigh.

The conclusion of this exercise was that the best site appeared to be Nuthampstead (another which, unlike Stansted, made the eventual Roskill short list) – only nine miles from Stansted, across the border in Hertfordshire. Armed with this evidence Crosland made a first and unsuccessful attempt to get the matter reopened in Cabinet.

As it happened, Nuthampstead rated badly in the final Roskill assessment. But the work done by Heath and his colleagues in the Economic Services Division did, nonetheless, have a lasting effect on the choice of the third London airport. As the Report of the Commission notes, their preparatory work, including a long memorandum on the methods which might be used in the Commission's search for a solution, the order in which the work might be tackled, and the role which a research team could play in the exercise, were 'of immense value' in the early stages of the Commission's work, though once the Commission had played

9

itself in, the Division 'discreetly disappeared from the Commission's ken and played no further part'.

The arrival of Crosland, with his willingness to look afresh at what most senior people in the Board of Trade had regarded, with relief, as an irrevocable decision, appears to have had a liberating effect on those within the Department who had long lamented its official attachment to Stansted – an effect enhanced by the subsequent arrival as Crosland's adviser of the economist Professor Alan Day. But for most of their colleagues it was a deeply depressing development.

It is easy to understand the chagrin with which the Board of Trade officials greeted the announcement that the work on which they had spent so much time was now to be challenged and possibly dispensed with altogether. The decision is taken; the book is closed; and one moves on to the next task. Then suddenly the decision is no longer safe; the book is reopened; and people who have moved forward to the next task are brought back to the old one. The pattern of departmental work is sharply disrupted.

But when the responsibility for the decision is then taken from the Department and given to some outside body – as it was soon to be with the setting-up of the Roskill Commission – the blow to confidence and harmony must be that much greater. Not only is there an implied slight on the Department's competence; there is also the depressing knowledge, that, at the end of the day, when the outsiders have come and gone, the responsibility for carrying out their handiwork, which in the Department's view may be much inferior to its own, will fall on them. In the case of Stansted, the reopening was all the more hurtful because the decision had evolved over such a long period, and so many in the Department had devoted much of their working lives to it.

It is simpler for an in-coming Minister to do this kind of thing to his staff than it is for a Minister who has already played his way in. No carefully-built bank of trust stands to be sacrificed. What is more, a Minister who goes back on a decision he has announced, or long defended, knows that the consequences may be embarrassing and painful for him. There will be the remonstrations of those who have been assured that it is a firm and irrevocable decision, and have laid their own plans on that basis. There will be criticism, even ridicule, both in Parliament and in the Press, where a politician's change of mind is often treated as if it were a symptom of some kind of moral degradation – unless, of course, the change has been specifically demanded in the paper's own editorial columns. There is also the fear that going back on a decision will be taken by political friends as well as political opponents as a sign of weakness. Here, it will perhaps be said, is a man who can be swayed if you press him hard enough. Such talk can lead to a sizeable dent in one's political reputation. Circumstances like these, impossible to measure and only possible to guess at when one is examining an individual decision, do nevertheless help to explain the tenacity with which a Minister and his Department will sometimes stand by a decision which seems, to many outside, to be patently wrong and certain to be overturned in the end.

And, though Crosland's decision to reopen the Stansted case was clearly an important step towards its eventual abandonment, one could not have been sure at that stage that anything would happen in the end. The re-examination could have been inconclusive. The Cabinet might have decided that it would be better to soldier on. But there was a more immediate difficulty. It is the tradition in British government that a decision once settled in Cabinet shall not be taken back to Cabinet unless there is some substantial new reason for reopening it. The situation must be shown to have changed in some fundamental way; and it was impossible to say that this had happened in the matter of the third London airport. The first move to reopen the affair during Crosland's re-examination foundered on precisely this ground.

It is, of course, a perfectly reasonable rule. There is more business for Cabinet to transact than they can easily cope with, especially at a time when crises are piling in as thick and fast as they were on the Labour Government towards the end of 1967.

So for the moment, the official line remained that Stansted would be built. News of what was happening within the Board of Trade nevertheless began to find its way into the papers. The theory that Heathrow would be saturated by the end of 1973, which had inspired the arguments about urgency brought forward by the BAA and accepted by the Board of Trade, was, it was reported, now in question. Heathrow's capacity was now thought to be greater than had been assumed; and the switch to jumbo jets was coming faster than the White Paper of May 1967 had allowed for. But officially, nothing was admitted. On 10 November, the *Daily Telegraph* reported Richard Crossman, who, as Leader of the House, was responsible for the conduct of the Government's business, as saying that Stansted would go ahead, and that the Development Order would come before Parliament 'soon'. On 11 December, Crosland's Private Secretary wrote to the Secretary of the Eastern Regional Advisory Committee of the TUC, which had been urging the merits of a site in the Brecklands area of Norfolk, or possibly a development of the base at Lakenheath, saying that in the Government's view neither of these choices was practicable and adding that they were satisfied that Stansted was the right choice. This line was to be maintained in the Commons as late as 7 February, when the Minister of State at the Board of Trade, J. P. W. Mallalieu, said at Question Time that he could confirm that there would be no further inquiries.

By this time, however, help had appeared from an unexpected quarter. A debate in the House of Lords on 11 December – the very day that Crosland's private secretary had sent his depressing message to the Eastern Regional Advisory Committee of the TUC – had demonstrated a near-unanimous condemnation of the Stansted decision (indeed, Lord Boothby, who had served thirty-four years in the Commons and had been a member of the Lords for nine, declared that he had never heard a more devastating debate in the whole of his parliamentary life). Though the Lords already had a pro-Foulness group, about forty strong, the debate showed that deeply entrenched feelings against what the Government were doing existed well outside that group. The clear impression left by the debate was that it would take a fearful struggle to get the Development Order through the House of Lords, and that it was even entirely possible that they would take the constitutionally very serious step of voting it down.

Nor did Ministers' efforts to defend the decision improve their chances of assuaging this general hostility. Indeed, the debate was opened by Lord Kennet with a speech which later led some of his admirers to speculate that it might have been a subtle attempt to sabotage the Government's chances of success. It began with one of those astonishing improvisations which had marked the government evidence at the Blake inquiry. Lord Kennet was explaining why it was that, on receiving the Blake Report, the Government had decided on a reappraisal behind closed doors rather than one in the open. He explained that:

> The Government were faced with the following dilemma. If the review were to be conducted in public, with public submissions from public bodies and private individuals, it would not be able to examine the defence problem in detail for obvious security reasons. If on the other hand, it were to make a proper appraisal of the defence aspect it would have to be done in private and as we all know private in-quiries carry less conviction than public ones.
>
> They decided that the defence aspect was so important – that is, that the right choice depended so much on how many defence airfields would have to be closed or reduced according to where you put your third London airport – that in spite of the disadvantages this review should be con-ducted in private.

These assertions were greeted with natural astonishment by many who heard them, particularly by those of them who had read the White Paper around which the debate was taking place. For the White Paper had contained a paragraph[3] explaining the reasons for holding the inquiry in private. This overriding need for security was not numbered among them; and various voices were now heard demanding anxiously to know why this should be. To which Lord Kennet ingenuously replied that there were always problems of packing all you wanted to say into any White Paper and this one had been excluded because of shortage of space; an assertion all the harder to believe since the White Paper had run to only thirty-four pages.

Lord Kennet's summoning up of the spirit of security only served to confirm the suspicion of many in the anti-Stansted camp that, not for the first time, the impenetrable secrets of the Ministry of Defence were being used as a convenient alibi to justify a decision which was otherwise inexplicable; and having experienced this kind of manoeuvre in debates on a variety of subjects before, they were reluctant to attach much weight to its invocation in the context of the third London airport. However, there were some factors in the Stansted decision which could be spoken about in decent society, and Lord Kennet had now moved on to discussing them. The cost of Stansted, for instance, was less than that of the other canvassed sites – although it was noted that the gap between Stansted and its rivals appeared to have narrowed still further since the figures had last been given. Much of the other argument, however, was totally unchanged. Not everyone was opposed to an airport on his doorstep: 'We must get out of our heads any suggestion that people, or a majority of people, will not or do not want to live close to airports. Certainly the noise is bad, but the work and prosperity produced is good.' Whitehall was best placed to weigh up all the factors; and though the sequence of consultation and inquiry had left room for improvement, even a better system would have led to Stansted.

Lord Dilhorne, a former Conservative Lord Chancellor, was weighty and cutting. The process of consultation had not only been inadequate; it looked to him as though it had been dangerously unconstitutional. 'Everyone would consider it monstrous that after a case had been heard, judgment should be based on matters which had not come out at the trial. Yet is not that what has happened here?'

Lord Walston, who had left the Board of Trade on the same day as Douglas Jay, still stood by Stansted, though he thought the whole story had been a 'sadly mismanaged affair'. Lord Sainsbury was sure the protestors were unrepresentative of public opinion as a whole. But mostly it was relentlessly one-way traffic. Twenty-four of the thirty speakers were totally opposed to the Stansted development. The demand by the leader of the Conservative peers, Lord Jellicoe, for a full comprehensive inquiry with the fullest terms of reference clearly reflected the view of a large section of the House.

There is no doubt that the evidence of this debate was the crucial factor which enabled Anthony Crosland to get the Stansted question back to Cabinet and to recruit some of those who had previously been classed as waverers to get a majority. Patrick Gordon Walker, who had succeeded Crosland at Education and was one of the Cabinet Ministers who took that decision, explained later in his book, *The Cabinet*[4]:

> The Cabinet found it very hard to reach a decision and considered the matter at length at three or four meetings. The conclusion was a reluctant one forced upon Ministers by what appeared to be unanswerable arguments.
>
> Publication of the decision to locate the third London airport at Stansted evoked considerable public concern and controversy. The Cabinet then discussed the issue again and amended its original decision in regard to the siting of the runways.
>
> The necessary Order was passed by the House of Commons[5] but the Cabinet became aware that it might be rejected in the Lords. Rejection of an Order by the Lords was extremely rare and, normally, would have caused no concern; it could easily be put right by another vote in the Commons. On this occasion, public opinion seemed to be stirred to a point that would make difficult a reversal of the Lords' rejection of the Order.

Two questions have to be asked about the effect of the Lords debate on the Cabinet's view of Stansted. Firstly, why should the

Lords have been ready to defy the Government on Stansted, and risk inevitable retribution in an imposed reform of the upper chamber, when on a series of other issues, they had accepted Government measures they disliked rather than bring down this wrath upon themselves? And secondly, why was the Government ready to accept the roadblock which an unrepresentative chamber (far less socially representative of the community at large than the local anti-airport lobby, whose middle-class air had created such grave Ministerial suspicion) had erected in their path?

For the Lords to vote down an Order was a very grave affair. Though their powers over legislation had been reduced to the right to delay, their power over delegated legislation had never been circumscribed in the same way. It was – as both main parties recognised – an anomaly; and it was bound to be dealt with in the coming round of Lords reform. The fact remained, however, that an Order defeated in the Lords would stay defeated. It could not be reinstated in the Commons. The only course was to start all over again.

The issue came to a head the following year when the Lords voted down an Order on Rhodesian sanctions. Immediately they got the punishment they must have known was coming to them; Harold Wilson called off talks on an agreed programme of Lords reform and announced that there would be an imposed solution. Clearly, had the Government decided to force through the Stansted Order, the consequences of a Lords refusal to accept it would have been no less serious. They would have been saving Stansted but directly sacrificing themselves. The calculation on the Conservative benches, however, was rather different from that over Rhodesia. On Stansted, there were two good reasons for thinking that, if confrontations were ever going to occur, there could be no better case than this one. Firstly, the Lords were generally agreed to have a special expertise on the environment; and secondly, they could reasonably assume (as they could not over Rhodesia) that any defiance of the Government would earn them wide public support. A government imposing reform on the

Lords after a rebellion on Rhodesia would have felt very confident about what it was doing. A government imposing reform on
a chamber which had just 'saved' Stansted might well have faced
an outcry.

For the Cabinet, the effect of the ambush being constructed in
the Lords was to make a policy whose main attraction had been
its speed and convenience – Stansted, some senior Ministers had
come to think, might not live up to the claims that were made for
it, but at least it could be built in time to cope with the increase
in demand – simple and convenient no longer; it would only be
achieved through a struggle with the Lords which would be a
drain on valuable time and on energies and attention more
urgently needed elsewhere. Again, as Gordon Walker says, the
extent to which the Lords, unrepresentative elite as they were,
seemed to reflect a growing body of outside opinion, could not
be discounted. It is one thing, a Minister of that day observed
later, to take on the Lords over an issue you are sure about; it
is quite another when you are not at all convinced you are
right. It was the appearance of the challenge from the Lords
on top of all the other accumulated evidence, from stickers on
car windows to cost benefit analysis in the Board of Trade,
which wrenched the door open. The Lords revolt, reflected one
Minister afterwards, 'simply cut along a line which was perforated
already'.

As it happened, another substantial pair of scissors could be
heard snapping away not far behind. The allegations of constitutional malpractice had reached the ears of the Council on
Tribunals, a watchdog body set up in 1958 as a result of recommendations by the Franks Committee.

It was a fairly rare occurrence for the Council to seize on an
issue, as it did now, and warn the Government about its behaviour.
This was only the third such intervention in the Council's history. The first and most famous of its incursions had been set off
by another Essex planning dispute. This concerned a planning
application (also, ironically, at Stansted) to quarry for chalk,
which was refused by the local planning authority on the grounds

that crops and livestock on neighbouring land would be affected. The Inspector at the subsequent inquiry agreed with this finding, but the Minister overruled him and granted planning permission. The Minister, it transpired, had set in hand further consultations with the Ministry of Agriculture after the local inquiry which had convinced him that the threat to livestock and crops was not as serious as evidence to the inquiry had suggested.

The obvious objection to this procedure, of course, was that the evidence which had settled the decision had never been examined before the inquiry. The demands of fair play suggested that the verdict in these matters ought to be settled on the basis of what the two sides said before the inquiry, and not as a result of backstairs conversations later on. It looked, too, as if the conduct of the Minister offended against the recommendation of the Franks Committee that, in cases of this kind, objectors should be informed of the new evidence and should be given the right to have the inquiry reopened – though this was not as clear as it might have been, since Franks applied this rule only to factual evidence and not to 'expert assistance in the evaluation of technical evidence given at the inquiry', and it was really very difficult to know where one of these activities stopped and the other began. The objectors put the facts before the Council on Tribunals, and as a result of subsequent discussions a clearer code of practice was worked out which was brought into force in 1962.

Had the Stansted inquiry been a statutory inquiry, there is no doubt at all that the Government's behaviour in the Stansted case would gravely have contravened the code. The Stansted Working Party had already submitted that the gap between the facts put before the Blake inquiry and the facts stated in the subsequent White Paper was wide enough to impose a moral duty on the Government to reopen the case. But at the end of 1968, a circumstance had arisen which made that argument still more compelling.

This immediately followed the failure of Crosland's first attempt to reopen the case. It was agreed that a concession should be made to local opposition by re-aligning the airport runways so as to diminish the noise effects over Bishop's Stortford, Harlow, Sawbridgeworth and to a lesser extent Thaxted. The decision was announced on 17 November and a letter explaining it was sent to local authorities in the new year by the local government minister, Anthony Greenwood. The tenor of the letter suggested that ministers sincerely felt the concession to be a generous one and expected congratulations; but if this was so, they were to be disappointed. The trouble was that the scheme now showed four runways (the target revealed in the White Paper) instead of two. That, of course, was bound to increase the cost of the project (in a television appearance on 12 December, the day after the Lords debate, Greenwood put the extra cost at £10 million). But it also meant that many people who would not have been directly affected by the old Stansted plan, as outlined before the inquiry, were bound to be affected by the new ones; while others who had been expecting to suffer a little from it now found that the threat was greater than before. The Stansted Working Party wrote immediately to Greenwood drawing the obvious conclusion; common justice required that people in these categories should now be given their chance to object before a reopened inquiry.

Now since the Stansted inquiry was not a statutory inquiry, the Government could, had it been sufficiently cynical, have replied to these charges simply by saying 'bad luck'. After all, they had not been under any legal obligation to hold any inquiry at all, so they could hardly be under any legal obligation now to institute a new one.

It happened, though, that the terms of reference of the Council on Tribunals had been changed in 1966, and they now decided to assert their right to lecture the Minister on his moral as well as on his statutory duty.[6] This was an entirely untested assertion of their new authority and one which it was certainly open to the Government to challenge.[7] The Council, under the chairmanship of a Labour peer, Baroness Burton of Coventry, decided it was worth the attempt. So they sent a special report to the Lord Chancellor asking that 'all those whose lives and property will be seriously affected should be given the opportunity of a fair hearing of their objections at a public inquiry'.

There is some conflict of opinion as to whether or not this declaration influenced the Government's decision to reopen the case. The extreme theory, which appeared in the *Sunday Times* and *Observer* the following weekend, was that the re-alignment of the runways had been a deliberate attempt by a Minister or Ministers to force the issue of backstairs consultation, alert the official watchdog, and so ensure the reopening of the case. Ronald Butt, in the *Sunday Times* attributed this plot to Crosland, while Nora Beloff in the *Observer* detected the hand of that master of the Machiavellian stratagem, Richard Crossman.

Ministers asserted at the time, and still do, that this imputation of fiendish cunning was more than they deserved. Newspapers next day reported in unattributable unanimity that the Council on Tribunals report had come too late to turn the issue. On the other hand, Richard Crossman, in trouble at a meeting of the Parliamentary Labour Party on the night of the reopening, was reported to have drawn attention to the insistence of the Council on Tribunals as to the course the Government must take.

Certainly the re-alignment of the runways was the chosen pretext for reopening the case when Crosland made his announcement in the Commons on 22 February. The decision was explained, however, as one made by Ministers after cool consideration of private rights and public duties – certainly not as a reflex action set off by the hot breath of the Council on Tribunals on their necks.

Replies to the Ministry of Housing's January letter had now come in and had been studied, Crosland said. They had shown that the re-alignment of runways would affect many who had had neither occasion nor opportunity to object at the previous inquiry. It had further been shown that the additional requirement for land, and the provision for expansion to four runways, were such a radical departure from the proposals previously published as to constitute virtually a new project.

In the light of these considerations [Crosland announced] and the continued public concern about the Stansted decision, and the general welcome given to the proposals in the Town and Country Planning Bill for dealing with cases of this type, the Government have decided to refer to an inquiry, which will take place in public, the question of the siting of the third London airport. Meanwhile, we shall not proceed with the Special Development Order for Stansted.

In the subsequent mood of jubilation (not universal, but clearly shared by a very many members), and amidst the mutual congratulations of the Conservative and Labour Members who had fought the Stansted proposal, one or two exchanges took place which were to assume some significance later on. Would the inquiry, asked Terry Boston (Lab., Faversham) – already taking up a defensive position on behalf of people in north Kent who would be threatened by several of the now fashionable proposals – examine the possibility of going outside the South-East? 'I do not want the terms of reference to be too restrictive,' Crosland replied, 'but almost everybody who has considered the problem is convinced that we shall need somewhere in the South-East another airport whatever we might need in other parts of the country in future years.' (The Liberal Member for Orpington, Eric Lubbock, promptly objected that *he* had considered the problem and hadn't been convinced of anything of the sort.)

A few Labour MPs were quick to signal their unwillingness to join in the celebrations. One north Kent MP, James Wellbeloved (Erith) spoke of 'a shameful act'. The veteran Manny Shinwell was also showing signs of displeasure. When the Labour MP for Tottenham, Norman Atkinson, welcomed the Government's decision, Shinwell objected that Atkinson had voted in favour of the choice of Stansted. 'I was at that time,' Atkinson excused himself, 'a rather reluctant prisoner.'

The moment of reckoning came later in the meeting of the Parliamentary Party where a succession of MPs thundered out their resentment and humiliation at the way they had been treated. Out of loyalty to the Party, they had voted in favour of a policy which most of them knew had been botched; and now here was the Government coolly abandoning it. The Government, said Charles Pannell, a former Minister who sat for Leeds West, had pushed through the decision on the payroll vote. (Harold Wilson's administration was the largest on record, and his critics frequently liked to say that this was due less to the search for the most comprehensive possible efficiency than to his desire to tie the hands of possible rebels.)

The Government had treated its back-benchers as lobby fodder, said a second back-bencher. The Government had publicly demonstrated its incompetence, said a third. Albert Murray (Gravesend) called it government by pressure group; even the Scottish Nationalists were now pushing the Government about. Perhaps the most poignant complaint was that of Eric Moonman (Billericay), who had a direct constituency interest in the affair. Essex MPs had subjugated their natural instincts to oppose the policy out of loyalty to the Party. Now they had been made to look ridiculous. The Stansted inquest, reported Ian Aitken in the *Guardian*, had caused more trouble at this meeting even than the dispute over the shutting of the door on the floods of Kenyan immigrants who were trying to get into the country – an issue which had troubled the consciences of almost every Labour MP.

But the recriminations within the Party were a small price to pay for having disposed at last of the albatross which Stansted had now become for every politician who came within distance of it. Not that the Stansted project was necessarily dead; as Crossman had protested, in his attempts to placate the angry PLP, while nobody could be proud of the history of Stansted, it was not true to say it was now being abandoned. (It could yet be chosen by the new commission.) If it was eventually to be chosen, however, at least the government of the day would be able to announce its decision without opening itself up to the wealth of impressive and valid condemnation which had been mustered against the reports of the two inter-departmental committees and against a variety of government witnesses at Chelmsford. The Commission should at last, in other words, take the political sting out of the London airports issue.

Yet could anyone really expect, after all this, that Stansted would be heard of any more? It had come very badly out of the reviews within the Board of Trade. And surely now, the failure of Whitehall to find realistic alternative sites – a failure mostly attributed to the reluctance of officials in the Ministry of Aviation and Board of Trade to be parted from the views they had been clutching on to for something like fifteen years – would be replaced by an approach of greater enterprise and imagination. Lord Jellicoe, in the Lords debate, had recalled how, as a Housing and Local Government minister, he had taken part in a Lords debate on the future of Ullswater, armed with a brief which declared that, in the unanimous view of the Government's experts, there was no alternative source whatever from which Manchester might take its water. And yet, when the superb oratory of the late Lord Birkett had persuaded the Lords to reject that solution and the Government had been driven to abandon it, a variety of alternatives had miraculously appeared. 'The moral is,' Lord Jellicoe said, 'that expert opinion, however expert, especially entrenched expert opinion, must be very carefully weighed indeed.'

But it would all be different now. The establishment dogma had been exploded. Any number of schemes, for airports on estuaries, airports in bogs, airports a hundred miles away to be linked by methods of transit hardly yet even on the drawing board, were hurriedly fetched out and dusted down in eager anticipation. One expected at any moment to learn that plans were to be submitted once more to site the thing on the roof of Charing Cross Station, or to disinter the revolutionary schemes of the ill-fated Charles Frobisher.

But at least in the villages and hamlets of east Hertfordshire and north-west Essex, the campaigners of the past four years could now sleep a little more easily. Their purses were lighter for it, of course. As Peter Kirk said later in a letter to *The Times*, local people had had to raise something like £35,000 to defend their rights, while the two county councils involved had been faced with sums mounting to well over £100,000 – sums which inevitably of course had to fall on the ratepayers of those counties, including those of Bishop's Stortford, Sawbridgeworth, Saffron Walden and Stansted Mountfitchet, who had also contributed in good measure to the fund raising appeals.

In a way, of course, a lot of it had been rather fun. Those bold declarations of defiance on the back window; those rousing, urgent meetings in council chamber and village hall those ambitious plans laid over morning coffee, or cocktails at half-past six; above all, perhaps, that sense of living in a community where rich and poor worked together for a common cause, where the place you lived in became, not just a convenient location, but something to which you were committed to your very soul. A sad feeling of anti-climax began to settle on these places as the prospect of a Stansted airport began to recede.

But while Stansted slept the tranquil sleep of the potentially reprieved, there was apprehension in another part of Essex. The choices before the new Commission were endless; the final choice might well fall in some unconsidered place two hundred miles away. But for some who lived in the furthest corner of south-east Essex, there was, from the moment Crosland's statement had been completed, an uneasy, even anguished feeling, that the finger must now be pointing at them.

8

. . . In which Foulness becomes fashionable, Shoeburyness becomes movable, and Cublington pledges a fight to the death

The campaign to put the airport at Foulness had now gained substantial new encouragement. Embedded in the speech of Lord Beswick, who wound up for the Government in the Lords debate, was an admission the Foulness lobby found very significant. He did not, he said, believe that the Stansted decision had been influenced by the heavy cost of moving the Shoeburyness ranges. 'I am bound to say, speaking for myself, that I have never treated this one very seriously, though I have no doubt it will cost a great deal to move it to some other ground.'

Despite the reservation that he spoke for himself and not for the Government, the Foulness lobby deduced – and rightly – that the long reign of the defence veto was now coming to an end. The Ministry case had been accepted in full by the Hole Committee; and despite the subsequent scornful treatment of it by both the Inspector and the Assessor at Chelmsford, the Peterson Committee had swallowed it too. The first tentative Board of Trade reappraisal had respected it – but the further study ordered by Crosland had not. The effect of Beswick's statement was to convince many proponents of the Foulness scheme that the last barrier had been swept away.

It was noted with particular alacrity at the Essex County Council headquarters at Chelmsford. Essex had long been interested in the possibilities of Foulness. It was a county particularly vulnerable to those on the hunt for airport sites; Stansted had already been singled out, but Nuthampstead too would have affected a large swathe of Essex countryside, and there had been

10

146 A SADLY MISMANAGED AFFAIR

another suggestion discussed within the Board of Trade to build at a place called Matching Green, just outside Harlow. Seven of the eighteen sites short-listed by the Hole Committee had been in the county; eighteen of the twenty-nine sites which were to reach Roskill's 'medium list' were either in Essex or in parts of Hertfordshire so near that Essex could not possibly escape the effects.

It was therefore an advantage to the County Council to be able to offer a positive suggestion of its own. Not only did it show that the Council was not just being parochial and obstructive, setting its own local interest above the interests of the nation as a whole; it also served to draw the fire away from other sites which in the Council's view would be much more deleterious to the well-being of the county as a whole. Foulness was, as it were, a kind of burnt offering to assuage those who might otherwise light on something much more precious.

Whitehall had done its best to damp down this enthusiasm. A letter to Essex County Council in the summer of 1965 had declared that the airport would 'in no circumstances' be placed at Foulness, where only recently £2 million had been spent on improvements to the range. But in spite of discouragements, the interest was maintained, and Foulness gradually began to emerge not simply as an unfortunate sacrifice to the national need but as a positive good. The planning office at Chelmsford seems at times to have been swept along by romantic dreams of what the airport might do for the south-east sector of the county. By the time of the Roskill local hearings, the county planning officer, D. Jennings-Smith, had become positively euphoric: 'The existing and the new development together,' he said in evidence to the local hearing of the Commission, 'would raise the population of south-east Essex over a threshold from which a whole new range of economic and social facilities could eventually be provided, such as, perhaps, a teaching hospital, university, operatic and ballet companies.' So Foulness was not only to have an airport; there would, perhaps, be a Royal Foulness Ballet too.

Foulness is the place you come to if, driving east, you persist beyond Southend. Until the Havengore Bridge was built in the twenties, the only road to the island was a causeway called the Broomway, submerged at high tide. It is a strange, remote, tranquil, almost other-worldly kind of place. Standing out by the sea wall it is easy to feel you are at the very edge of England.

The Army has the use of most of it – the farmers, most of whom belong to families who have worked this land for generations, are tenants of the military, which has the right to invade their fields at any time. Driving out towards the sea, one is struck by that atmosphere of good natured scruffiness which often accompanies the Army; yet one can also sense what a calm, closed, dignified and silent place this must have been before the Army lighted on it in 1805 in their search for a suitable site for Colonel Shrapnel to practise his explosions.

To the north, off the island and across the River Roach, there are small informal scatterings of houses; none worth a mention in Pevsner, perhaps, but each part of the casual accumulated charm which Betjeman noted in the main street of Stewkley. The churches, too, do not compare in historic importance with the Norman church of Stewkley; but they have a rich engaging charm. The church at Foulness has a tower which leans at an endearing, almost apologetic, angle to it. There are other churches, like the one which lurks on a bend in the road near Stambridge, which have white wooden spires. The light off the sea has a sweetness which neither north Buckinghamshire nor inland Essex can match.

Little of this has been commemorated. There is none of the casual prettiness of the lanes and hamlets around Stansted, none of the cosy kemptness of the villages of north Bucks. It is not the sort of atmosphere you can capture on a postcard. 'You can't photograph tranquillity,' says Derrick Wood, the most persistent of the small group who have fought to save Foulness. You could paint it perhaps? 'Maybe. But Constable was too far north for us.'

Out at sea there are the Brent geese, who settle where they

find a particular kind of grass which grows in profusion here. Much time was devoted before Roskill to considering whether or not they would be a serious hazard to flying. Efforts have since been made to tempt them away, to a stretch of coast near Chichester where the grass could be specially grown, or perhaps to an island further down the Thames Estuary. But in the two years following Roskill, their numbers increased. When I went to Foulness, on a crisp morning in November 1972, they were massed out to sea like an invading army. 'My friend the naturalist,' said a Foulness man, looking out towards them, 'reckons there's enough of them now to stop this airport.'

The Noise Abatement Society ('they ought,' said a bitter campaigner against the Foulness project, 'to change their name to the Noise Transference Society') recommended Foulness as early as 1957, though at that stage it was conceived as an airport built inland. They continued to champion it throughout the deliberations over Stansted; at the Chelmsford inquiry, they gave evidence in favour of it, and called the honorary secretary of the Monorail Group of Great Britain to demonstrate that it could be 'brought within twenty minutes of London'. (The distance from London to Foulness as the crow flies is forty-six miles, though if you went there by road in 1972 it felt like twice as much.)

In November 1967, the NAS developed the argument for their Foulness proposals in their most sophisticated form so far. A publication containing reports by consultants on various aspects of the project – now planned for offshore construction – argued that Foulness was preferable to Stansted on air traffic grounds, on grounds of regional planning and on general considerations of amenity. And they calculated that it would cost only £7 million more than Stansted.

By this time, however, a new and extremely ambitious scheme for an airport at Foulness was being floated by a London consulting engineer, Bernard L. Clark. On 21 June 1967, the *Daily Telegraph* reported that Clark had completed a scheme to build one of the largest airports in the world on a man-made island in the Thames Estuary off Foulness, adding that it might make possible the closing of Heathrow. The *Telegraph* pursued this story vigorously throughout the summer and autumn. Clark himself conducted his campaign with bustling enthusiasm. 'We're the instigators of this thing,' he declared at the Roskill local inquiry at Southend. In a subsequent interview with the Southend *Evening Echo* (4 May 1971) he said he had spent six years and £50,000 working on the project, but according to Brian Cashinella and Keith Thompson in *Permission to Land*[1] (Cashinella wrote most of the original *Daily Telegraph* reports), 'the concept of developing the Thames Estuary into a gigantic industrial-residential complex had been mulling over in his mind since the 1930s.'

Clark's enthusiasm went beyond the drawing board and the *Daily Telegraph*. He began travelling up to south-east Essex and addressing public meetings. His theme was always the same; this was not only the imaginative project which Britain needed, it was also the opening up of untold wealth for the people of Southend and environs. As he put it somewhat more mildly before Roskill: 'I feel that somehow, with due respect, the people of Southend are virtually sitting on a veritable gold mine.' There was quite a lot of scepticism about Clark's operations. 'I thought he was a bit of a buffoon, but he was certainly persuasive,' says one man who attended several of his meetings. 'What I particularly remember is that he always had very good maps.'

On the whole, though, there was plenty of curiosity and some excitement in Southend about what Clark was saying. It is often an agreeable feeling to be told you are living in the Klondike of tomorrow. The prospect of an airport at Foulness, as told by Bernard Clark, became much more glamorous than it had been made to look in the more sober expositions of the Noise Abaters.

On the days of the debate on Stansted, the plush red benches of the House of Lords were scattered with batches of documents advocating the merits of Foulness. Few had come without their briefs from the NAS and from Bernard Clark; and many were eager to quote from them. At times the debate sounded almost like a public meeting organised by the Foulness Promotion Company. Enthusiasm was all the greater, one suspects from reading the report of the debate, because many of the speakers had no first-hand knowledge of the area about which they were talking. 'I understand,' said Lord Dilhorne of the Shoeburyness installation, 'that there is a very good officers' mess there. I gather the rest of it is pretty obsolete.' It was about the only institution in the area to come in for a word of commendation. The general impression seemed to be – in the words of Sir David Renton in a later Commons debate – that 'Foulness seems to have been aptly named by our forebears'. There were, by the end of the debate, all the necessary signs that a bandwagon had been launched on the road and was now careering away with a happy band of enthusiasts hanging out of its windows chanting a variety of tempting slogans.

What the bandwagon needed now was plenty of money to keep it in good running order, and that was not far behind. On 15 January – halfway between the Lords debate and the Crosland statement on the reopening – the *Daily Telegraph* led with a story reporting that Sir John Howard had formed a company to build an airport and deep-water docks.

This was the Thames Estuary Development Company, more details of which were to emerge before and during the Roskill hearings. Sir John himself headed a company with a long record of big construction projects; it had laid the foundations for the Forth and Severn bridges. He was a former chairman of the National Union of the Conservative Party, and lived near Thurleigh. (*The Times* reported on 5 April 1971 that Sir John had been a contributor to inland airport resistance funds, but had refused to tell them how much he had given.)

As outlined before the Roskill commission, TEDCO was composed of the following interests: the Port of London Authority (whose chairman, Lord Simon, became chairman of TEDCO), Southend Borough Council, John Howard and Co. Ltd., London and Thames Haven Oil Wharves Ltd., and John Mowlem and Co. Ltd.

This was a very shrewd assembly of interests. The presence of Southend was particularly significant for reasons which will emerge in a moment. But the Port of London Authority, government financed and run by a government-appointed board, seemed a particularly telling recruit.[2]

The PLA called a press conference on 13 February 1969, to explain its own thinking on the scheme. Airport and seaport were to be joined by an oil terminal. As for Shoeburyness, Sir John Howard was able to tell the press that they believed they had found another place where the Government could put it, though its location was not disclosed.

A second group now emerged expressing its willingness to take the cost of building the complex off the taxpayer's shoulders. This was the Thames Aeroport Group, led by David Thornton, a former Conservative Councillor on the Greater London Council. At the time of the local hearings, their members were said to include Holland, Hannen & Cubitts Ltd., the London Dredging Company Ltd., Limmer and Trinidad Co. Ltd., C. A. Blackwell (Contracts) Ltd., West's Piling and Construction Co. Ltd., General Asphalte Co. Ltd., Felnex Central Properties Ltd., and Slough Estates Ltd. Several more companies were said to be waiting to join. Lord Holford, a favourite choice as adviser to governments and local authorities on planning affairs, had been enrolled as a planning consultant. The TAG operation showed unmistakeable signs of haste. Their submissions to the Roskill Commission were a month late in arriving and the member of the Commission who took the local inquiries, A. J. Hunt, described himself as 'more than uneasy' about it.

The attempts of these groups to make a favourable impression on the Roskill Commission and so on the wider audience of public and politicians were to fall painfully flat. The Roskill Report frankly expresses deep suspicions about their motives and intentions, and in any case, Roskill did not think the kind of mammoth complex they envisaged (though there were fall-back plans for an airport alone) were at all desirable. It would not, the Commission concluded, be consistent with regional planning

strategy. This was not, of course, a matter in which the Commission could decide: 'it is not for us to say whether a seaport is
viable, desirable or technically feasible or preferable to one elsewhere.' Even so: 'there have been times during our work when
we have doubted whether some of those who pressed the claims
of a joint project have had the interests of the airport in the forefront of their minds.' They could see the attractions, of course;
the economics of the seaport project would be greatly enhanced
if the material dredged from the sea could be used for land re-
clamation, especially so close to the site. So if a seaport at Foulness
was the prime intention, an airport project beside it would be a
godsend. But was that the right way to approach the decision?
'The prospect of such commercial benefits to those reclaiming the
land will not necessarily encourage objective appraisal of the
merits of Foulness as an airport location.'

Roskill also suspected that the groups had, in their accountancy,
perhaps been carried away by their commercial enthusiasm.
'Quite apart from the different ways of estimating the possible
savings [TEDCO had put them at more than £30 million,
Roskill's own researchers at £15 million] there are considerable
doubts about whether a seaport project of the kind envisaged
would be undertaken at Foulness.' As for TEDCO's claim that
they could build an airport at no cost to the taxpayer, that was
greeted with withering sarcasm:

> We can well understand that such a scheme would be attrac-
> tive to its promoters as well as lenders, for the entire risk of
> failure of the airport would be borne by the authority and
> thus ultimately by the taxpayer. None would be borne by
> the promoters of the scheme. We have more difficulty in
> seeing what advantages there are to the taxpayer in such a
> scheme.

TEDCO in fact modified this well-advertised claim in the late
stages of the Commission but the Commission's misgivings were
not allayed.

But the political significance of the TEDCO–TAG submissions (and these were only the two which emerged as serious contenders; others had been drawn by the juicy bait but had not compiled submissions for the inquiry, while still others were holding their hand until they saw what the Government decided) was rather different. These interests failed to persuade Roskill; but he was not the final arbiter. Would the Government be tempted by possible savings of public money to overrule an inland site in favour of Foulness? Or, on a more cynical view, freely expressed by some political opponents, did they mean to choose Foulness in the hope of satisfying their business friends?

In the end, Foulness was indeed chosen; and the seaport was included with the airport. TEDCO and TAG, however, were – at least for the moment – rejected. Peter Walker, Secretary of State for the Environment, who by now had overall charge of the whole project, decided to keep control in the hands of government and government agencies to an extent which dismayed some of his own back-benchers. The one successful contender from the TEDCO–TAG stable was the Port of London Authority whom no one could class as a greedy capitalist profiteer.

But this does not by any means demonstrate that the intervention of powerful business interests made no difference in the end. Simply in terms of the publicity they commanded, together with their lobbying in Whitehall and Westminster, they helped to impress the desirability of the Foulness scheme on a much wider audience than might otherwise have been possible. They may have been excluded (at least at the time when this is being written) from any share of the pickings, but they must have had a hand in the final outcome.

The other lasting result of the TEDCO involvement was to line the Borough Council of Southend firmly up in the camp of the Foulness airport supporters. This could well have happened anyway; the argument that prosperity was round the corner had clearly appealed both to commercial interests and to ordinary people in the town. On the other hand, residents associations in places likely to be most affected – Shoeburyness and Thorpe Bay – were firmly opposed to the project.

The man chiefly responsible for this development was the Conservative Leader of the Council, Norman Harris. Harris was a formidable figure in local politics who had moved on to a wider stage with his appointment to the South-East Regional Planning Council. He was chairman and managing director of a group of engineering companies and an aviation enthusiast (he held a private pilot's licence). He was the sort of tirelessly active interventionist councillor who gets dubbed by his local newspaper with titles such as 'Mr Southend'.

The Borough's involvement in the consortium was made public at the PLA press conference on 13 February 1969. According to the *Southend Standard* it was the culmination of a long history of negotiations involving Harris and the Town Clerk, Archibald Glen. The *Standard* was delighted: 'when commercial interests no longer have the scope to point the way ahead,' it declared, 'Britain will be in a very poor shape indeed.'

The acumen of its Leader and Town Clerk was quickly demonstrated to the Council by the announcement on 4 March that Foulness was on the short list of the Roskill Commission. When the Council met on 27 March, therefore, the prospect of a Foulness airport seemed doubly desirable. Not only did many townspeople see the prospect of a new prosperity for the area; but they could also expect, thanks to municipal initiative, direct participation in the profits. The economic advantages of the project, declared one councillor, were so great that people who objected ought to move out of the area. The Council then passed by an 'overwhelming' majority a resolution welcoming the Foulness proposal subject to certain safeguards; these included freedom from 'intrusive noise' over the Borough, assurances that costs of the project should not fall on local ratepayers, and satisfactory terms to mitigate the effect on the municipal airport.

Some of those who took part in this debate complained afterwards that the resolution had been steam-rollered through. A Liberal councillor told the Roskill local hearing that the debate was all over in twenty-five minutes and that the resolution passed reflected the opinions not of the burgesses of Southend but of the

Conservative Party caucus. (The effect of this submission was perhaps a little diminished however by his subsequent admission that he had himself voted in favour.) In fact, the only opposition came from two councillors who unsuccessfully moved that the resolution be referred back. One other councillor who had voted against the proposal in the caucus meeting – he represented Shoeburyness – was absent. As he explained later at an angry meeting of his residents association: 'I have a private life.' 'When you get on to the Council,' he added, 'you have to toe the line on these issues.'

Since Southend (population 166,000) was by far the largest local authority in the area, and since Essex had been publicly advocating Foulness for several years, the Southend verdict removed the last hope of those who opposed the project that they might get the same kind of protection which Essex County Council had given to Stansted, and which the appropriate local authorities were now holding out to people in and around Cublington, Thurleigh and Nuthampstead. On the whole, though, the evidence suggests that this would have happened even without the TEDCO involvement. The Southend Labour Party approved the project and the Trades Council had joined forces with the Chamber of Trade in an organisation, perhaps unique in having such a composition, which, having been to see the Europoort developments across the Channel, was enthusiastically in favour of creating something similar off Southend. But the taking of the town into TEDCO – unknown to the townspeople until concluded, and announced before the Council voted on the Foulness project, must have swung the chance that Southend would oppose the airport from unlikely to impossible.[3]

So, well before the time that the Roskill short list was ready, Foulness was already distinct from any of the other sites in that its merits were being proclaimed by a powerful alliance of interests with a great deal of money at their command – while resistance was left to a few spontaneous groups with no political base and no resources at all. In this, the Foulness situation was the exact reverse of the state of the parties at the three inland sites.

It was, on the face of it, a somewhat lopsided contest. But that had been inevitable. Apart from the campaign for Foulness, the only sites which attracted clamorous support were all so far from London as to be certain non-runners. That was true of the East Midlands airport at Castle Donnington; it was even more true of the proposal to build on Thorne Waste, which, whatever its other merits, had the incurable disadvantage of being not far from Goole in Yorkshire.

It was one of the strengths of the Roskill Commission however, that, unlike the politicians, it was to be impervious to nobbling by pressure groups. That was one of the advantages of putting the choice before a forum of this kind. But the main reason why the Government decided to hand over the problem to Roskill was much simpler. The whole London airport story had become so littered by suspicion, confusion and the apparent breaking of pledges; by mismanagement, misjudgment and a shying away from unfortunate facts, that it was no longer likely that any solution which came from Whitehall would be accepted without question. If the Government had been prepared, now, to say that Foulness must be the choice, that might have been acceptable. But there was no evidence to justify that. Its advantages had been treated to massive publicity; its disadvantages, which on the most unsophisticated scrutiny could clearly be detected, were much less talked about. So a decision to go inland still seemed, looking at all considerations, to be very much on the cards. The people might not take it from the politicians; but they might take it from an impartial commission.

Roskill was not, as it has sometimes been called, a Royal Commission. It is by no means clear how these distinctions are made; some problems are assigned to Royal Commissions, some to commissions of more plebeian aspect, and some to simple committees. Those who look for logic in these dispositions tend to be disappointed;[4] some subjects studied by committees are clearly quite as important as others which rate a Royal Commission.

Roskill was considered to be very important; but it would report to the Minister not the Sovereign. That made little difference to its weight. The power to summon witnesses and send for papers is automatic with a Royal Commission, because of the Royal prerogative. But that is a power easily conferred on other commissions. Being a Royal Commission is like being a Royal Borough; it adds to prestige and looks good on the letterhead, but it doesn't add significantly to your power.

Assembling for a moment all these Royal Commissions, commissions, and committees into one huge amorphous parcel, one needs to make two kinds of distinction which go deeper than mere nomenclature. The first is the motive of the Government in setting them up; the second is the way that members are selected.

Sometimes a commission is appointed because a problem is causing great embarrassment but does not, in the Government's view, require urgent action. A commission is a way of passing the time to political advantage. The Crowther Commission on the Constitution, set up in 1968, was designed to demonstrate to the Welsh and Scots, who had been transferring their loyalties in alarming numbers from Labour to the Nationalist parties, that the Government knew and understood their aspirations and wished to take the highest advice on what should be done. The Government had correctly estimated, first, that it would take the Commission a long time to report (during which its existence provided the best possible excuse for inaction) and that anyway when the next General Election came round the Nationalists might have begun to go away.

Then there is the situation in which the immensity of a problem is perceived but there are genuine doubts as to what should be done about it. This was the case with Roskill; it has also been the reason for successive attempts to re-plan local government through the reports of committees and commissions.

And then there are commissions set up because the Government knows perfectly well that something needs doing but feels that the public is not ready for it – probably because it will cost the earth and the taxpayer will be alarmed unless he can be tranquillised first by the impartial verdict of a high authority. The Robbins Committee Report on higher education is an excellent example of this type. An even more blatant example is the committee set up by Richard Crossman in 1969 when he was Secretary of State for Social Services into the plight of unsupported mothers.

Crossman wrote subsequently of this decision:

> when we appointed the Finer Committee I assumed that it would make the obvious recommendation. In future the deserted mother would not have to depend on alimony but would draw the same benefit as the widow, which she would receive as of right from the DHSS. Having taken responsibility for her maintenance, the state would have the job of collecting cash from her husband.[5]

Why, if this was so obvious, did Crossman not simply take the decision himself? Because he calculated, quite accurately, that general consent did not exist for this change. Certain influential groups were likely to regard it with deep suspicion: the Conservative Party, for one; the Treasury, for another. How much more acceptable it might look if it came from the mouths of an impartial and well-qualified committee instead of by the decree of a notoriously socialist politician!

This leads directly to the second consideration, which is: how do you staff your commission? Anxious that the 'obvious' solution should not be overlooked, Crossman appointed as chairman a Labour-inclined, *New Statesman*-contributing lawyer, Morris Finer and sat beside him (among others) a *Daily Mirror* columnist, Marjorie Proops, and the woman's officer of the Transport and General Workers' Union, Marie Patterson. (This kind of tactic can only be judged by results. If you select a committee which is too clearly biased towards a given solution, you make it too easy later on for its findings to be discounted.)

A more sophisticated example of a commission which appears to have been deliberately pointed in a certain direction is the

Fulton Commission on the Civil Service. When Harold Wilson came to power, it was certain that his reforming eye would be set on the Civil Service. He was, after all, set on making Britain more modern, more rational and less subject to the rule of gentlemen; and anyone with this brief who believed one-tenth of what people told him would have wanted to reform the Civil Service. In any case, he had made his intentions pretty clear in a series of discussions with an Oxford specialist in these matters, Norman Hunt. In these talks, put out on radio and reprinted in the *Listener*, civil servants were able to read that the leader of the Labour Party thought there was too much amateurism about them and that both their institutions and their attitudes needed changing.

However, the point about the Civil Service is that it is impartial; not, as is sometimes supposed, between government and opposition, but impartial because it obliterates its natural repugnance to the advertised policy of any party which may form the government and concentrates (at least in theory) in serving an in-coming government of the Left with the same devotion as it served an out-going government of the Right; and vice versa. For a party politician, even if he *was* Prime Minister, to seek to impose his known views on it, to the extent of altering its whole character, was therefore unthinkable. The chosen Commission, however, included a reformist Labour MP, Shirley Williams (who, when she was appointed a Minister, was replaced by another of comparable tendency) and the chairman of a Fabian Group which had recommended a number of changes similar to those which eventually appeared in the Fulton Report. (Conservatism, meanwhile, was represented by Sir Edward Boyle.) On top of this, the same Norman Hunt who had sat at Wilson's feet in the *Listener* not only appeared on the committee but worked (unlike the rest) full time, and led its management consultancy team. Not surprisingly, the eventual report had a distinctly Wilsonian echo to it – although closer inspection showed that the ringing radicalism of its introductory chapter was not really matched by the relative moderation of what followed.

The Redcliffe-Maud Commission on Local Government

seemed to some observers to belong to this category also, especially perhaps because of its connection with Richard Crossman,
who set it up in his days as Minister of Housing and Local Government. The presence of Derek Senior, former planning correspondent of the *Guardian*, who, alone of those taking part, had written
at length on what kind of reform was necessary, made it likely
that he might be a governing influence on the report; though in
fact he failed to carry his colleagues with him and ended up
writing a dissenting report.

The crucial decision in many of these cases is whether it is
better to let interests centrally involved sit on the commission
or whether it is better for them to appear as witnesses. Senior
could have been an influential witness to the Maud Commission
but instead was enrolled as a member. The key decision in setting
up the Donovan Commission was what you did about the TUC.
In the event, the Labour Government, recognising perhaps its
especial vulnerability in this area, gave a place to its general
secretary, George Woodcock, thus improving the chances of a
report acceptable to the unions, while reducing the likelihood
that it would advocate anything radically different from the status
quo. The equivalent with Roskill might have been to put the head
of the Airports Authority, Peter Masefield, on the Commission.

It will be seen, therefore, that the selection of a committee
or commission is a fertile field for wangling and wrangling.
The more there is at stake, the more bitter the fighting is likely
to be. The final responsibility for the choice will lie with the
Minister or Ministers responsible (if the committee is sufficiently
important, the Prime Minister himself will demand the final
word). But the whole process will be subject to consultations of
the most comprehensive and wearisome kind; for setting up a
commission is a little like sending out party invitations; the last
person you are likely to think of could well be offended by
exclusion and go into a long and even damaging sulk.

The selection of the Roskill Commission was no exception.
The first and most difficult job was to choose a chairman, and
that meant first deciding what sort of chairman – administrator,

businessman, academic or judge – you wanted him to be. Because they were determined to extract the political sting from the issue, the Government had long talks with Conservative representatives – the Deputy-Leader, Reginald Maudling; aviation spokesman Frederick (later Sir Frederick) Corfield; Chief Whip William Whitelaw; and the former Chancellor, Lord Dilhorne, who had been a powerful critic of the Government in the Lords. The Conservatives, especially Dilhorne, wanted a judge to head the inquiry. Crosland equally definitely did not. In the end, though, the Government conceded. There were still grave reservations about having what would inevitably seem like a legal tribunal; and as subsequent events were to show, these were by no means misplaced. But Opposition consent to the enterprise was essential, and the Government decided to yield.

When you have chosen the chairman, the selection of members can begin. Here the Government was insistent on representing a wide range of interests, and the final selection was much in line with the immediate tentative indications which Crosland gave to the Commons when he announced the terms of reference.[6] The members were: Alfred Goldstein, partner in the consulting engineers, Messrs R. Travers Morgan and Partners, and a seasoned server on official committees (Goldstein was to emerge as one of the most formidable members of the team with a firm grasp of the whole of the complicated field it was surveying and an ability in questioning witnesses which any lawyer might have envied); A. J. Hunt, Planning Inspector, the Department of the Environment (Hunt had the sensitive and wearing task of conducting the local inquiries); David Keith-Lucas, Professor of Aircraft Design at the Cranfield Institute of Technology; Arthur Knight, who served as the 'statutory businessman' – almost as regular a component of commissions as the more celebrated 'statutory woman'; Alan Walters, Cassel Professor of Economics in London University and an exponent of the econometric techniques which were to become a crucial part of the Commission's operations; and Colin Buchanan, Professor of Transport, Imperial College of Science and Technology.

Whitehall maintains a list of people suitable and likely to be available for work of this kind – 'the great and good' as they are familiarly called. But it is only there as a fallback. None of these names seems to have been drawn directly from it. Most were recommendations by civil servants and departmental advisers. Crosland's own instinct was to get economic expertise on the Commission. As it transpired, he now had a chairman with no known interest in sophisticated techniques of economic measurement; but this would be more than offset by the high-powered research operation whose merits were now being successfully impressed on the Commissioners.

The one name which leapt out of the page when the list was published was that of Buchanan. There appears to have been some resistance to Buchanan at the Board of Trade end of the operation and he was by no means an automatic choice. But the Ministry of Housing, now taking a much more active part in fighting such battles than they had seven years before, wanted a formidable watchdog for their interests and Buchanan had a track record to justify such a role.

He was best known, of course, as the leader of the group which had produced the report *Traffic in Towns*. But his experience of official committees went back longer than that. As the Ministry of Housing man on a working party on London traffic in the early fifties, he had found himself part of a team on which representatives of the motoring organisations, the motor trade and the commercial vehicle operators had sat with civil servants, and to which the unrestricted march of the motor vehicle had apparently seemed no bad thing. They proposed widespread building of car parks under squares. This would, it was admitted, mean the destruction of trees, but even this was justified; trees were bound to die some time anyway; squares had originally been laid out without trees; and anyway they had changed so much since the eighteenth century that there had already been 'an appreciable change in amenities'. Buchanan put in a minority report, boldly asserting the heretical view that squares were too valuable to be destroyed and attacking the idea, otherwise unanimously held by the Committee, that one ought to subsidise long-term parkers.[7] Given this courageous (some said stubborn) streak, and given too the approach he had brought to *Traffic in Towns*, Buchanan's likely attitudes to the choice of the third London airport, his overwhelming concern with the environmental aspects of the decision, could have been forecast.[8]

The bias towards the expertise of the economist which Crosland rated as so important was one to which Buchanan was unsympathetic from the very beginning. After it was all over, Buchanan said in a public lecture[9] – reported by Tony Aldous in his book, *Battle for the Environment*[10] – that the Commission had probably got off to a wrong start. Though this was 'a planning matter if ever there was one' it had been set up under the auspices of the Board of Trade (airports) rather than the Ministry of Housing (planning) and had therefore been 'heavily weighted on the economic side'.

The research team were not all economists. The group also included people with planning, transport and engineering expertise. In size, scope and resources they constituted the most sophisticated research organisation yet attached to any commission, and they operated in a much more public way than other research groups had done. The research team became in the later stages of Roskill almost like a group of outside participants in the inquiry; they gave their own evidence, and they had, like everyone else, to justify their calculations. The forecasts they put forward could be assessed and compared with what other forecasters had said – with the Commission free to prefer other people's calculations in the light of the debate to those of its own staff. Other commissions of the time, like Donovan on labour relations and Redcliffe-Maud on local government, were able to call on and commission research; but none had an independent task force at its disposal to match that given to Roskill, and none was able to submit the evidence of its researchers to such continued and rigorous challenge.

But the final guarantee that the whole question should be subjected to the closest economic analysis was assured by the place given to cost benefit analysis in the Commission's working methods. (Their terms of reference enjoined them to consider its use but did not instruct them to use it.) Although cost benefit analysis was later to come in for violent criticism, even to be elevated, in a speech early in 1973 from that prophet of the scientific and professional, Harold Wilson, to the status of a public enemy, it appeared at the time Roskill was set up in a very different light. It was even hailed as the people's friend; especially when, by widening the conventional cost calculations and bringing in factors of social convenience which few would have thought of, Christopher Foster used it to justify the building of the Victoria line. The Stansted campaigners had clamoured for it, on the grounds that if you set the losses to the environment and the lives of the inhabitants against the simple aviation construction and transport calculations, Stansted would have become a good deal less attractive than before, and other places, Sheppey perhaps, substantially more desirable.

But as the Roskill team went about their calculations, the environmental lobby began to change its view. Certainly you could use cost benefit analysis to widen the field of your calculation. But you then met the difficulty that some of the things which seemed most valuable to you wouldn't fit within the technique. They could not be quantified, even by a research team as resourceful as Roskill's. The research team, faced with Stewkley's Norman church, was, as crime reporters used to say about Scotland Yard, baffled. Here was one of three Norman churches in the country which had not been substantially added to, the usual Victorian urge to improve having been mercifully resisted by G. E. Street. What tests could you use to value that? At first, they used the fire insurance figure, which was £51,000; and when this was greeted with derision, they removed the figure altogether – which was more realistic, but meant that Stewkley church didn't figure at all in the final calculation.

The truth about cost benefit analysis is that, like a computer, it is only as good as what you feed into it. There are things it cannot do. If you ask a computer to sing, it is no use complaining that you do not like the result. If you ask cost benefit analysis to put a figure on a Norman church, it will inevitably retire defeated.

Now in theory this defect should by no means have proved fatal. Cost benefit analysis is an aid to rational decision making, like a slide rule or an Ordnance Survey map. It will *help* you work towards the right answers but it will not *find* them for you. But in practice if you mangle your input or if you let these figures, incomplete as they inevitably are, dominate your operation, you are lost.

As time went on Roskill's cost benefit analysis was subjected to growing criticism. Some thought it mumbo jumbo, pretending to a scientific status it did not deserve; 'nonsense on stilts', in the words of one of the most celebrated assaults on it. Others queried the way it was used,[11] or objected to the value put on individual components. But the really central question was this: was it true, as critics alleged, that Roskill, like the sorcerer's apprentice, had set in hand a process he could not stop; that instead of being the tool of the Commission, cost benefit analysis had become almost the master? The Commission devoted much space in their Report to rejecting this conclusion; they had never pretended that cost benefit analysis told the whole of the story; they had merely weighed its evidence against all the rest which had come before them. But they had difficulty in being believed. The cost benefit analysis had 'found' for Cublington and against Foulness. It had done so largely because of one item, which told heavily against Foulness – the cost of journey time to the airport – and this, as it happened, was the most contentious calculation in the whole edifice. They had then written a Report which said Cublington was to be preferred to Foulness, a contention which many people in government departments, rich country houses, humble cottages, the offices of great newspapers, churches and chapels, clubs and pubs and coffee bars were firmly convinced was nonsensical. The conclusion was obvious; Roskill had become imprisoned in his cost benefit analysis, and the whole Report had been biased accordingly.

Increasingly, the Cublington lobby came to see cost benefit analysis as a dangerous weapon which was unmistakeably pointing at them. So they – and the other inland groups – began to do their best to discredit and diminish it.[12] The arguments which Buchanan was to use in his dissenting report had been commonplace among the Cublington witnesses. Fred Pooley, the lively and imaginative County Architect and Planning Officer of Buckinghamshire County Council, argued it at the local inquiry:

> Instinctive judgments should not be lightly set aside be-cause of the findings of various studies which are so often inconclusive and arrive at the wrong answer. In any case they can never put a true value on things like a Beethoven symphony, a man on the moon, or the value of the retention of a piece of satisfactory environment, and they would – if they had been in vogue in the eighteenth century – have demanded that the High at Oxford and the Crescent at Bath should both have been built on straight lines.

The extent to which Roskill employed these techniques was not implicit in the idea of the inquiry, but it was clearly affected by the new thinking within the Board of Trade which had gained fresh impetus since Crosland's arrival, and it left its mark on the Commission. But the eventual results of Roskill were also – as the results of all such inquiries must be – shaped by the terms of reference.

The job of choosing the terms of reference belongs to White-hall, and if they turn out to be, in the eyes of objectors, grossly restrictive and unfair – as they had been at Gatwick – or simply somewhat inadequate, as they had been in the Stansted inquiry, then that is bad luck, and it is no use appealing because there is no one you can appeal to. If a government rigs the terms of reference too much in its own favour, then there is likely to be an outcry, and it may become a Party matter, which will undermine the whole exercise and therefore be counter-productive. Un-fortunately, however, the fact that terms of reference are not what most people would want them to be cannot necessarily be guaranteed to alert an opposition. The nearer the inquiry comes to Westminster and Whitehall, the tighter the terms of reference are likely to be. The most objectionable instance of this was the Younger Committee's inquiry into privacy, set up in 1970, which could be as radical as it liked about private intrusion into personal freedom but was barred from uttering the slightest peep of criticism about the behaviour of Whitehall; that, it was cosily explained, was the subject of another investigation – by a private Whitehall committee. Appallingly inadequate though its brief turned out to be, the Committee agreed to work on it, though its eventual Report was punctuated by groans of frustration at the artificial restrictions which had been placed upon its operations.

In the case of the Roskill inquiry, the terms of reference produced little immediate complaint. They were:

> To inquire into the timing of the need for a four-runway airport to cater for the growth of traffic at existing airports serving the London area, to consider the various alternative sites, and to recommend which site should be selected.

In addition, the attention of the Commission was also specifically directed to general planning issues including population and employment growth, noise, amenity, and effect on agriculture and existing property; aviation issues, including air traffic control and safety; surface access; defence issues; and cost, including the need for cost benefit analysis.

In announcing this on 20 May 1968, Crosland added:

> The form of the Inquiry must meet two requirements. On the one hand, this is one of the most important investment and planning decisions which the nation must make in the next decade. This points to an expert, rigorous and systematic study of the many and complex problems involved. At the same time, the decision will profoundly affect the lives of thousands of people living near the chosen site. This calls for an adequate method of representation of the local interests involved.

All this, like the choice of chairman, had been agreed between Government and Opposition, but the precise wording of the terms of reference and Crosland's explanation have some political importance because they were later to be blamed for the alleged 'failure' of the Roskill Commission. It was argued both that the terms of reference were wrong, and that Roskill had failed to respond to them.

The chief complaint about the framing of the terms was that they assumed the need for a four-runway airport, and assumed also that it must be built in the South-East. Whatever may be said with hindsight, Crosland was justified in asserting that few – allowing of course for mavericks like Eric Lubbock[13] – had questioned either principle. London, as Peter Masefield, chairman of the British Airports Authority, was fond of saying, was the 'honeypot'. What evidence there was supported him. Origin and destination surveys in 1968 and 1969 showed that something like seventy per cent of all British non-business passengers originated from south-east England; and Peter Masefield had shown that seventy-five per cent of tourists who came to this country spent no nights anywhere other than in London. There was much more; none of it ideal (the origin and destination surveys had been conducted exclusively in the South-East); but such as it was, that is what it said.

Crosland was later to argue, however, that the terms of reference were framed in such a way that Roskill could have reported that no airport in the South-East was needed; or alternatively that a two-runway site would do. In a Commons debate on the Roskill Report on 4 March 1971 he declared that they had made a mistake of 'immense' significance in the way they had read the terms of reference. (He did not add, though others were busy pointing it out, that, perhaps uniquely among such Commissions, Roskill failed to reproduce its terms of reference within its Report.)

The terms of reference had put timing first, selection of sites afterwards. Suppose Roskill had done the job in that order; his study of timing might have shown that Heathrow and Gatwick –

perhaps with additional use of other existing sites – could cope with all traffic in the foreseeable future. If the terms of reference were to blame, said Crosland, then he admitted his share of the guilt. Probably it had been a mistake specifically to refer to a four-runway airport in the opening sentence 'though we did so in good faith and with good motives, because at the time almost all expert opinion was stressing the urgency of the need and assumed that there was a need for a four-runway airport'.

It seemed, on the face of it, rather harsh to criticise Roskill for sharing that general assumption. But Crosland went on to point out a supplementary answer he had later given which he thought should have clarified the position.[14] It would be open to the Commission, he had said, to say that there was no need for a third airport or that it should be indefinitely postponed.

The sense of frustration and futility which this criticism of Crosland's must have created in anyone who had wrestled with airports policy over the years, and most of all in anyone who had served on, researched for, or testified to the Roskill Commission is almost too much to bear. Here was the man who had set up the inquiry asserting that its labours, the million pounds or more it had cost, and a great part of the time and money it had cost the objectors, might have been so much waste.

In fact, the possibility of writing into the terms of reference a specific instruction to consider whether any new airport was needed at all does seem to have been considered by Crosland and his advisers, some of whom opposed it, successfully, as almost an act of provocation. And undoubtedly the question was confronted in the early stages of the Roskill Commission, though with something of a sense of awe; for any adequate assessment of the chances of doing without the airport would have demanded an even wider, more searching, and more time-consuming inquiry than that on which they were already embarked. Had Crosland decided to put this idea before the Commission in the form of a sentence in the terms of reference, rather than leaving a supplementary answer in the Commons to carry the weight, the Commission's response might have been different. At any rate, the Commission made its own judgment, and there was no loud protest at the time. No one sought, and no one gave, the hard and fast ruling which would have removed all ambiguities.

After the terms of reference, the next point to be established is the way in which a commission should carry out its inquiries. The Roskill Commission's procedure was partly derived from those new and sophisticated kinds of planning inquiry which had been promised by Ministers during the Stansted debates and which had subsequently been embodied in the Town and Country Planning Act of 1968. (The White Paper had been published the day before the Commons debate on Stansted.) But it was necessarily adjusted to meet the special need of the third London airport situation; it was never intended that planning inquiries under the new system should be presided over by a High Court judge.

The announcement of the Commission's chairman was made on 20 May; his colleagues were named on 24 June and the Commission met for the first time next day. The first round of public hearings, at which major issues were examined in general terms, began on 1 November. Written submissions were also considered at this stage, and the trawl of sites (which produced a long list of seventy-eight, some as distant as Newark and Leicester) was carried out by two specially commissioned teams.

All this evidence, plus the work to that date of the research team, was discussed at a meeting at Exeter College, Oxford (where Roskill had been an undergraduate some forty years before) the weekend before Christmas. There were already signs of strain in the operation, and these became clearer as the 'medium list' of twenty-nine was whittled down to four. Already there was a philosophical conflict between Buchanan and the majority; and the issue was Foulness. By the time the medium list had been pruned once more to produce a 'reduced medium list' it was clear that Foulness was not, on the system of assessments which had been agreed, a very serious contender. The running order at this stage was headed by Nuthampstead, with Hockliffe (just east of Leighton Buzzard) in second place, Silverstone third, Cublington fourth and Thurleigh fifth. Foulness was in thirteenth place in a list of fifteen – still behind Stansted, which was ninth. However you looked at the figures, Foulness didn't look capable of making the short list.

So the logical course was to leave it out; and some of the Commission were ready to do it. Politically, however (for even the Roskill Commission couldn't overlook considerations of *internal* politics) that was a dangerous course; for it became clear that if Foulness went, Buchanan was likely to go too.

But there was an argument, quite apart from the possible loss of Buchanan – which would clearly have wrecked the Commission's chances of producing an authoritative report – for squeezing Foulness in. One needed a benchmark against which the inland sites could be rated. The comparison between the inland sites and the most favoured of the estuarial sites seemed essential to an effective report.

This was the background to a letter written by Roskill to Crosland on 24 February 1969 – almost exactly a year after the announcement of the setting up of the Commission – to explain the short list which had been compiled. Foulness, it said, had come 'relatively low' on the cost benefit analysis so far. Yet the Commission believed that it might offer 'substantial advantages' over the inland sites, not least on regional planning grounds. The Commission had therefore decided to include it on the short list. The letter added: 'You will have noticed that Stansted does not find a place upon the short list, while Nuthampstead had been included. In the Commission's view, of the possible sites in that area, Nuthampstead offers most advantages over Stansted, particularly in respect of noise and air traffic control compatibility with Heathrow.'

The short list (Cublington, Foulness, Nuthampstead, Thurleigh) was published on 4 March. The immediate reaction seemed to be satisfaction that Foulness had emerged as a real contender, and an air of 'told you so' over the elimination of Stansted. (The one notable abstainer from this cry was Douglas Jay, who remarked that Stansted hadn't really been excluded; it was still on the list, but they were now calling it Nuthampstead.)

The one complete newcomer on this list was Cublington. Nuthampstead and Thurleigh had come up in the Board of Trade review – though this was not known at the time. Thurleigh had also been floated, though unenthusiastically, by Peter Masefield in an interview in the magazine *The Director* in September 1967. Cublington seems first to have been mentioned by Arthur Reed, Air Correspondent of *The Times*, on 22 February – less than a fortnight before the short list appeared.

The assumption in north Buckinghamshire had been that the real threat was Silverstone. An organisation called the Silverstone Airport Resistance Association had already been formed. Silverstone is twelve miles from Cublington. Just as many of those who had campaigned against Stansted were removing their 'Say No to Stansted' stickers on the night of 4 March, only to make room for stickers saying 'Say No to Nuthampstead', so most of those who had been preparing to defend themselves against a threat from Silverstone now found that the danger was coming from the south of them rather than from the north. So the Silverstone Association simply switched around. It became instead the Wing Airport Resistance Association – for though Roskill talked of Cublington, those who knew the scruffy former Second World War airport the Commission had selected, now occupied by a chicken farm, knew it as Wing.

Each of the four sites now became the centre of a thriving 'hands off' industry. But the dynamics of Foulness's situation were, as has already been suggested, the reverse of the others. Foulness was actually being promoted, not only by a virtually unanimous alliance of the people in the threatened inland areas, but by local people as well. At Foulness, it was the anti-airport group which was the weak minority, lacking political organisation and resources; at the other three sites, as earlier at Stansted, it was the advocates of the airport who took on that role.

The unbridgeable gap between the strength of the defenders of Cublington, Nuthampstead and Thurleigh and the weakness of the defenders of Foulness was dramatically demonstrated by the series of local hearings conducted by A. J. Hunt between the beginning of May and the end of September.

Foulness came first – though with characteristic courtesy the Commission put back the date of the hearing when there were complaints of lack of time to prepare submissions. The Foulness hearing opened at the Civic Centre, Southend on Monday, 5 May. The evidence took eight days.

An informal alliance of parish councils, residents' groups and private individuals challenged the airport both on the traditional scores which arose in every inquiry – noise, urbanisation, the loss of agricultural land, loss of amenity and the rest – but also on an accumulation of objections peculiar to this area. The island community – deep-rooted, interrelated so that almost everyone you met was the first or second cousin of everyone else – would be completely destroyed. There was serious danger of flooding; there was a danger that the airport might sink (objects on the sand, said the Foulness islanders, tended to get swiftly swallowed; there had been one particularly gruesome case of a big timber ship which vanished in a twinkling). Noise was a worse problem here than in many places, because of its high sound transmission; this had been noted by Samuel Pepys, while more recently, yachtsmen engaged in private conversation had been overheard a full half mile away. (That, no doubt, was a warning to be passed on to Edward Heath, soon to become Prime Minister, who himself went sailing in these waters.) Then there were local industries which would be obliterated. The Leigh-on-Sea Shellfish Merchants Association felt bound to point out that the balance of payments would suffer as imports would be needed, in the face of the insatiable national demand for cockles, to replace their supplies. The whiteweed industry testified too and said Britain would be driven to fall back on inferior stuff from Germany.

But this opposition was hopelessly fragmented. A stream of local authorities queued up with Essex and Southend to welcome the airport into their midst – or at least, to declare that they would not positively oppose it. They included the Borough Council of Maldon; the Development Corporation of Basildon; the Urban Councils of Basildon, Benfleet, Canvey Island and Rayleigh, all represented by the same counsel as Essex; the Rural Councils of Rochford, Maldon and Tendring; and the councils of several of the parishes which would be most closely affected. The village of Paglesham might be pledged to oppose the airport; but the village of Canewdon, a couple of miles down the road, wanted it to come. There was even to be found, testifying at the local hearings in favour of the Foulness site, a breeder of geese; they'd emigrate to the Broadwater and Colne, he forecast, just as they did when the Shoeburyness guns were firing.

There had been hopes that, having been abandoned by Essex, the protestors might get some protection from Kent. After all, Kent, as Roskill was himself to point out in his Report, got the worst of both worlds: the noise of aeroplanes over its coast, without the compensating advantage of an airport close at hand (for it would mean a long detour, up one side of the estuary and down the other); and without the equally compelling compensation of new jobs in an area of inadequate employment. But Kent had decided not to oppose Foulness, only to signal its alarm about possible noise and to plead for this to be mitigated. Kent took the attitude that Hertfordshire's planning committee had tried to take over Stansted: 'it lies outside Kent and the County Council take the view that it can properly concern itself only with any consequences which might flow from the selection which directly affect Kent.'

Alone of the big battalions, the BAA was on the protestors' side. The Authority did not condemn the project in so many words but it did not disguise, either before Roskill or elsewhere, its misgivings. None of the sites was good; but Foulness was decidedly the worst. The Authority's annual reports reflected gloomily on the challenge of the new French airport at Roissy, so close to the centre of the capital.

The worst blow for the objectors, however, came when the bulwark which had protected them so long was revealed at last to have crumbled quietly away. The defence objection to Shoeburyness was dead and buried by the time of the Roskill inquiry. What really mattered now to the Ministry, it appeared, was the base at Brize Norton.

Like the Inspector and Assessor at the Stansted inquiry, the Commission were not impressed by the long series of declarations which had led the Hole Committee to regard Shoeburyness as 'irreplaceable', which had inspired the warning off of Essex County Council in 1965 on the grounds that moving the range was 'out of the question', and had driven the Peterson Committee to conclude that the practicability of shifting it was 'not worth pursuing'. They calmly decided to ignore this objection in compiling the short list. And sure enough, when the Ministry of Defence later reported on its reappraisal of Shoeburyness – set in hand at Roskill's behest early in the inquiry – the finding was that Shoeburyness need not be regarded as an immovable obstacle – 'a fact', comments the Report tartly, 'which we had firmly assumed when compiling the short list, notwithstanding the contrary views which had been so strongly expressed on this matter in the past.' Indeed, the Ministry now asserted, in secret session with the Commission, that Foulness, so long unthinkable, was in defence terms actually the best of the four short-listed sites.

Having prised this juicy bone from the Ministry's mouth, the Commission aimed some well-deserved kicks in its direction. In the light of what had happened over Shoeburyness, it was clearly not very much moved by the Ministry's newly acquired devotion to Brize Norton. (The Ministry had said that it would not wish to veto the choice of Cublington on this score but had issued what was 'tantamount to a warning that a recommendation of Cublington could encounter opposition'.) 'As we listened to the evidence,' says the Report, in perhaps the best of the superbly feline sentences which Roskill permitted himself now and then, 'we wondered whether Brize Norton was not in danger of becoming the Shoeburyness of Cublington.' They went on to assess the arguments on Brize Norton with what they described as 'doubt' but sounded rather more like total disbelief. And they clearly regarded the Ministry's figure for the cost of shifting establishments as hardly more realistic than those given at the time of the Stansted inquiry. Most ironic of all, the Commission pointed out that on present Ministry of Defence thinking, Nuthampstead would mean closing Shoeburyness too; so much for the argument that Stansted, only nine miles from Nuthampstead, was a desirable site because it would allow Shoeburyness to continue. Their one obvious concession to the Ministry was to recommend that if Shoeburyness were to be moved, the thing should be done quickly. (At the end of 1972, there was still no sign of a replacement site.)

The pattern of the Southend local hearing reinforced earlier indications that in political terms Foulness was a very convenient solution. The three local hearings which followed further underscored that conclusion. The Nuthampstead hearings, which ran from seven days from 9 June at Royston in Hertfordshire, produced the predictable clutch of protesting local authorities, headed by the 'Essex group' – Essex, Hertfordshire, Cambridgeshire and the Isle of Ely County Councils backed by town and gown interests from Cambridge, by the Nuthampstead Preservation Association and by private objectors.

Of the twenty-seven parties who appeared or were represented, only one individual submission favoured the airport. But in any case, the history of the Nuthampstead hearings is of purely antiquarian interest, for the objections which emerged to this proposal were so overwhelming that it was never at any time a serious contender for the crown.

The Thurleigh hearings, which ran for only four days (8 to 11 September) were more contentious. The parade of protesting local authorities, led here by the 'Bedfordshire' group of County Councils – Bedfordshire, Huntingdon and Peterborough, Northamptonshire, Buckinghamshire, Cambridgeshire and Isle of Ely – was supported by a well-organised grass-roots organisation called BARA (Bedfordshire Airport Resistance Association), which claimed 5,800 members. It had been first discussed when rumours began to appear that Thurleigh was on the short list and had held its inaugural meeting on 12 March. Though nowhere near so demonstrative as their counterparts in WARA, the members who attended the hearings at Bedford Corn Exchange were often vociferous: 'this is not Old Trafford', Hunt rebuked them on one occasion.

Unlike Nuthampstead, however, the Thurleigh proposal was not treated to almost universal condemnation. An analysis of votes taken in parish councils and at parish meetings showed some had expressed no opposition (though none had expressed jubilant acclaim). Those who had supposed feudalism to be dead in Britain might have noted with alarm that several of the villages which did not oppose the airport were on the Duke of Bedford's estate; the Duke himself was reported to be sympathetic to the project.

But the split in local opinion appeared to go deeper. Unlike Nuthampstead, Thurleigh appeared to have sprouted a fervent pro-airport group – and one with a sharp nose for national publicity. This was TECDA – the Thurleigh Emergency Committee for Democratic Action – and its message was similar to that of SAPA at Stansted: the anti-airport protestors, it said, are not representative. They are the articulate middle class. There is a great body of working-class opinion which would welcome the airport and the work which it would bring with it.

TECDA deserves some attention, for in the subsequent political equations the existence of a sturdy assembly of airport supporters was to become quite significant. It comprised, the local hearing was told, the Huntingdon Constituency Labour Party and fifteen local trade union branches, of which seven belonged to the AUEFW. Three more branches were said to have joined during the inquiry. Its principal spokesman was James Curran, a twenty-four-year-old Cambridge undergraduate who was chairman of the constituency party.

Curran argued his case with great enthusiasm and some skill before the inquiry, playing continually on the middle-class bias of the other side, emphasising the economic backwardness[15] of the area and suggesting that the concept of human happiness, of which much had been heard from BARA in its defence of quiet

12

lives and sleepy villages, might possibly be taken to include the chance of a good job. Against Curran's wholehearted young idealist, counsel for BARA, Hugh Forbes, seemed to have been cast in the role of heavy father. 'Young Mr Curran's attempt to bring politics into this particular area', he said of TECDA; as though politics, like the jumbo jet, was inimical to the whole spirit of the place.

Young Mr Curran's case did, however, look rather less substantial when Forbes had finished with it than it did at the start of the hearings. It transpired that TECDA owed its origins, not to any spontaneous coming together of the unions which belonged to it, but to a piece of private enterprise by Curran and a friend. They had simply written to every union branch they could locate asking if they would support a Thurleigh airport. The branches which were listed were those which had responded favourably, but Forbes insisted – and the point was not disproved – that in many cases the support consisted of a favourable reply by a branch official, possibly with no meeting or mandate from the branch. The meeting of Huntingdon Labour Party had been attended by only twelve people who had divided nine to three in favour of the airport; the prospective Labour candidate for the constituency, Martin Lawn,[16] opposed the TECDA campaign.

The fact that this support for the Thurleigh proposal had been drummed up by Curran's intervention does not, of course, mean that it did not deserve to be taken seriously. Indeed Curran's performance at the final hearings impressed the Commission and led directly to changes in the cost benefit calculations. What needs to be said, however, is that people who were not at the local hearings were to be heard asserting later that 'a lot of people at Thurleigh wanted it to go there' – a view certainly held by some of the politicians who were closely involved with the airports decision and which may have been based on assumptions about the status of TECDA which reflected Curran's talent for publicity rather than the actual substance of his organisation.

The most passionate, even riotous, of the local hearings, however, was that at Aylesbury from 14 to 24 July with eight days spent taking evidence. The scoreboard compiled by Hunt shows the way opposition had exploded at Cublington to an extent quite unmatched at the other inland sites – though it should be said that this was partly a reflection of organisation; the number of submissions at the Aylesbury hearing was inflated to a ridiculous and even potentially counter-productive degree by a campaign to get every local objector to turn in his own submission.

	Parties given leave to appear	Parties appearing	Written submissions	Percentage pro-airport
Cublington	818	203	725*	$2\frac{1}{2}$
Thurleigh	64	33	178	2
Nuthampstead	64	27	166	nil
Foulness	76	49	80	7

Source: Roskill Report, Appendix 4, Report of Mr A. J. Hunt, a member of the Commission, upon the Local Hearings.

But the Cublington campaign differed in kind as well as quantity. The bitter passion with which the case was fought, and the frequently[17] vindictive things said about the proposal, were unmatched elsewhere. Derrick Wood, at Southend, had agreed that for all his deep-rooted opposition to the Foulness proposal, he would be prepared to accept it if that was what the Commission recommended. But when Robert Maxwell, Labour MP for Buckinghamshire, made a similar admission at Aylesbury he was booed by the public gallery.

The northern reaches of Buckinghamshire, from which this resistance principally sprang, look, by and large, as though they have done rather well out of life. The core of the area which was threatened lies east of the agreeable open road which runs north-west out of Aylesbury; the sort of place where you can settle

* Over 400 of these were attributed to the letter-writing campaign.

back and feel free of the last tentacles of London, out in the still unsullied countryside. It is peopled with small and apparently quite prosperous settlements – some, like Whitchurch on the Buckingham road, with the air of small towns about them, independent and self-contained; others, like Cublington itself, just a scatter of farms and houses.

Three villages were to have been surrendered to the airport. One, Dunton, had two farms, a church (whose rector thundered against 'sacrilege') and little more. Cublington was perhaps two hundred people at the crossing of two minor roads. The most substantial casualty would have been Stewkley, immediately to the north of the present airfield.

Stewkley (population 1,250) is the kind of place one could drive through without noting that anything particular had happened. What immediately strikes one about it is its straggle; it is strung out for well over a mile along the road from Winslow to Wing, with side-turnings petering out abruptly as if the place suffered (which it assuredly does not) from some form of anaemia. You could hardly mistake the significance of the church, with its Norman tower and decorated arches; but of the rest, the most one might recall is the occasional entertaining house in the long medley of architectural styles which makes up the village street. To call it beautiful would be a sentimental exaggeration – not even Betjeman did that. But it is easier to see why some of the villagers claimed it to be unique.

If one were compiling a treasury of English villages, none of these three would be likely to get in. But some of the others, threatened not with destruction but by noise and by urbanisation which would slowly have destroyed their essential nature, are full of that unpremeditated charm which cannot be rebuilt or replaced; Weedon, with its fine domestic houses clustered round the bend on the road, set on a small hill; Oving, full of enticing looking lanes and twittens; North Marston, with its cluttered houses and eccentric pattern of streets, its imposing church presiding over all.

The full glory of this countryside, said its defenders at the

Roskill hearings, is in its wider landscapes. The best of the re-commended views is said to be from Muswell Hill – not the London suburb, but a promontory near the Oxfordshire border at Brill. But Ivinghoe Beacon does pretty well and may be more convenient. If you climb the Beacon, denying yourself the view till you reach the very top, and then turn back to look across the vale, you see what Buchanan meant. The less committed observer will perhaps be more conscious of the blemishes; to enjoy this panorama of unspoiled England you must first be able to close your mind to the ugly polluting presence of a cement works in the near foreground. But once you have done that, it is easy to imagine the vale suddenly filled with the angry roar of jet planes, dominating the whole of this calm and peaceable scene.

As anyone who has ever been up in an aeroplane knows, England is not the mass of almost unbroken urban growth which the arguments of the extreme environmental lobby would some-times seem to suggest. This part of north Buckinghamshire is a wide unbroken landscape, predominantly green. But the edge of London is only thirty miles to the south; the edge of the Midlands conurbation is only forty miles to the north-west. A break in the urban landscape is all the more precious here, on the axis of Britain's prosperity and growth.

The villages here claimed to have a deep-rootedness and social ease about them which would make their destruction, or their radical alteration by the economic pressures created by an airport, more serious than in the areas which were also under threat. That claim was sustained by a research project carried out by the University of Essex for the Roskill Commission.[18] It showed a higher degree of kinship, more local allegiances in terms of com-munity activity, and – whatever might subsequently be said about the class construction of the Cublington campaign – a more working-class 'profile' than Nuthampstead or Thurleigh.

The voice of this community was WARA – the former Silver-stone resistance group which had turned south against the Commission's unexpected strike. Stewkley maintained its own resistance organisation; they were, of course, the most violently threatened, but the people of Stewkley also, according to its schoolmaster, Geoffrey Ginn, who became its principal spokesman, possessed 'talents and fighting spirit' which were recognised by surrounding villages, so that Stewkley had emerged as 'the natural leader' of the anti-airport campaign.

This part of Buckinghamshire might be able to display a 'more working-class profile' than its rivals; but it had other useful advantages too. WARA, the Commission was told, had attracted 'all walks of life'. Some, though, were from more rarefied walks than others. It is clearly an advantage, if you are setting up a grassroots organisation of this kind, to be able to install a Rothschild as your treasurer.[19] It is also of advantage to have, newly arrived in your midst, the proprietor of a great national newspaper and his wife; the presence of Lady Hartwell on the executive committee was unlikely to diminish the zeal with which the *Daily Telegraph* covered the Cublington resistance. It is an advantage, too, if you have, already active in your field, amenity groups containing important and influential figures: the committee of the Friends of the Vale of Aylesbury, for example, included a senior Minister in the 1970 Conservative Government, Lord Carrington, as well as the Speaker of the House of Commons, Selwyn Lloyd, and the erstwhile scourge of Stansted, Douglas Jay.

The attack mounted on the Cublington proposal at the local hearings was the most formidable of the series. The institutional witnesses were led by the 'Buckinghamshire group' – Buckinghamshire, Oxfordshire and Northamptonshire – backed by a comprehensive march-past of the other interested counties (for Bedfordshire, like Buckinghamshire, had both Cublington and Thurleigh to fight). They had an influential ally in the development corporation of the new town of Milton Keynes. If Brize Norton was the Shoeburyness of Cublington, as Roskill suggested, then, remembering the Gatwick story, one might equally say that Milton Keynes was Cublington's Crawley; a new town, hardly set in hand before an unwanted airport was to be parked on its doorstep.

Page upon page was set before the local hearing setting out the implacable opposition of almost every parish in the area.[20] All these were cunningly itemised to rebut the inevitable accusations of middle- and upper-class bias; the occupations of parish councillors were listed, showing the retired railway workers along with the farmers, the welders with the London stockbrokers, the lowly cottager alongside the squire, or his lady, from the Hall. The petition, too, signed by more than 61,000 people, had been professionally audited; no chance here that people would sign twice, or that signatures would appear casting doubt on the whole enterprise, as the submissions of the Chartists had been marred by the appearance among supporting signatures of one which read 'Victoria Rex'. It was a petition – another advantage here for inland sites against those on the seaboard – signed by people from many counties, even from other lands; for the Vale of Aylesbury, unlike Foulness Island, is a place many people have driven through on the way to somewhere.

The protestors emphasised the threatened destruction not just of the physical but of the social character of the place. The consultants talked in terms of the broad sweep, the men in the street in terms of the cottage door. Tom Hancock, the planning consultant to WARA, told the local hearing: 'I hold that the combination of the strong "Natural growth" corridor together with the powerful catalyst of the third London airport would produce a very rapid growth, which would completely urbanise the sub-region and pre-empt any rational regional planning.' And in an echo of Lewis Keeble at the Chelmsford inquiry (if you want growth in this area, simply light the fuse and retire) Hancock added: 'Siting an airport at Cublington would be putting a match to a powder keg.'

Beside this bold and sweeping statement on regional, even national priorities, WARA set the plaintive appeals of those whose interest was merely parochial, confined to a single village, even a single street. The villages eagerly proclaimed their own special merits. 'Description of the people,' said the evidence submitted by Winslow. 'Exceptionally friendly, and quickly embrace newcomers into a wide variety of organisations. Atmosphere of general goodwill and fellowship. Wide cross-section of livelihoods and walks of life who have formed community with unusual degree of camaraderie.' 'An area of friendship,' wrote a resident of Stewkley about his village. 'Here have I found and known more close companionship that I could have done in a lifetime at Kingston-upon-Thames.'

There were frequent appeals to unchallengeable outside authority. The Queen was enlisted by a succession of witnesses:

> in her Commonwealth Day message [wrote E. G. R. Taylor] Her Majesty the Queen stated that she was reassured that human beings are more important than machines and that the happiness of mankind is even more important than the spectacular achievements of science. We of the inland sites implore the Roskill Commission to note well Her Majesty's statement.

Even higher authority was summoned to the defence of Cublington. St Mark and St Paul were quoted with such freedom that one might have expected them to be ushered at any moment on to the witness stand; while one or two who testified invoked an authority even more awesome than theirs. The Rector of Dunton, for instance, the Reverend Hubert Sillitoe:

> I may be mocked for being emotional – a common charge when faced with naked and embarrassing truth – but of course there is emotion. For we are all men and women; without emotion we become robots, which it is the evident ambition of some to make us. God *is* – *is*. He lives and reigns and he will have the last word – overriding the courts of justice in the day of judgment which, as the Son of God warned, 'with the same measure that we mete to others it will be measured to you again . . .' This will be no storm in a teacup but a violent hurricane, inspired by the spirit of God which will possess the souls of men. We shall if necessary fight – fight to the bitter end – for our homes, the graves of our dead ones, and our House of God. [Applause.]

It was all very different from the calm and sensible air of those hearings in Southend two months before. The sustained emotion of some of the Cublington hearings had, to some who were not caught up in it, an intimidating quality about it. The Rector of Dunton was cheered with the same fervour as the more moderate Robert Maxwell had been booed. Even the courteous and patient Hunt showed occasional signs of distress. There was, too, detectable in some of these hearings, a sense of a wider attack; on the bureaucratic enemy, for its constant incursion on people's freedom to do as they chose; or on the socialists who were behind this attack on middle-class England. When Hunt reminded protestors that the Commission had been 'appointed by the Government to do a task' there were cries of 'What a government?'

And it was very nearly all one-way traffic. The Southend hearings had some real debate about them; local people, local interests, debated the issues between them; points raised on one side were taken up on the other. But Aylesbury was almost a procession. The protestors had the floor almost to themselves; by now the main opposition to them appeared to be the absent research team.

The exception was CASC – the Cublington Airport Supporters Committee. But they were weak and hopelessly under-organised. They would anyway have been desperately vulnerable to the legal might arraigned against them; but they made matters worse by asserting claims they were quite unable to sustain.

They said they had 600 members, but declined at first to produce evidence to prove it on the grounds that their records were confidential, though they later agreed to submit the names in writing to the Commission. (CASC complained bitterly of intimidation by opponents of the airport and the Commission responded with a warning that this kind of thing must stop.) They claimed that the Duke of Bedford was their President, but were unable to prove this either. (Would it really have mattered, though, either way?)

They were courageous to the point of foolhardiness. Their

chairman, Edward Payne, wrote to the *Bucks Advertiser* the month before the local inquiry:

> I am sick and tired of hearing numerous little 'know-it-alls' standing up and criticising the Roskill Commission's preliminary report on the third London airport site at Cublington. Surely the Roskill Commission are a group of highly professional men who know their job and not a crowd of morons as the anti-airport fighters make them out to be. There is already too much disrespect and ridicule-making of authority as it is in this country. My advice to local 'would-be Roskill Commissioners' is to leave the job to the genuine one chosen by HM Government.

The Inspector could have been forgiven if he had sometimes imagined that he must have been transported to the judgment seat of Mr Justice Cocklecarrot, and that the hearing might at any moment be invaded by Miss Ruby Staggage and Mrs Rhoda Tasker,[21] doing battle in the case of the twelve red-bearded dwarfs. There was an interminable story about an article in the *Lancet* which allegedly said that the rate of apparent nervous breakdown in north Bucks was the highest in the country, the implication being that even an international airport could hardly make things worse, and would at least alleviate the prevailing boredom. Grave efforts were made to track down this story, the staff of the *Lancet* having been asked and having denied all knowledge of it. The source of the information was solemnly traced from witness to witness until the inquiry lighted at last on a Mrs Walsh, who said she had read it in her dentist's waiting room. But had she a copy? 'We are not,' she replied imperiously, 'in the habit of removing magazines from the dentist's surgery.' CASC's efforts were further impeded by the appearance of a local farmer who was first said to be a vice-president, though it was later asserted equally firmly that he was not. His plan was to build an underground railway from London to Manchester via Northampton, Birmingham and Leicester – a scheme which would link ideally, he pointed out, with an airport at Wing. 'I have experienced this feeling of elation with noise,' he declared while cross-examining, 'and I like to sing at the top of my voice while operating machinery.' 'Most interesting,' sighed the weary Inspector, 'but it is not a question.'

Here and there the bones of a serious case emerged; there were too few jobs in the area and not enough recreation; communications and services were running down and needed an economic stimulus like the airport to revive them. But CASC was too spontaneous and untutored an organisation to make much of an impression, and the framework of the Roskill inquiry, in which professionalism paid so well, reduced them to farce.

This, it should be said, was hardly the fault of Hunt, whose handling of the hearings, as reported in the transcript, leaves one full of admiration. But even the local hearings, relatively informal as they were in comparison with the sittings of the full Commission, were an intimidating experience for the weaker interests which appeared before them. 'I have attended a number of meetings,' a local councillor, giving evidence for TECDA at Bedford, suddenly burst out in the middle of cross-examination, 'but the actual legal standing of this meeting, and the expertise you are fighting against – it is overwhelming.' It is one of the most deeply unsatisfactory features of the whole Roskill hearings that a number of minority witnesses should have been so clearly overawed. Perhaps the one man who held his own with the serried ranks of barristry and consultancy was the leading Foulness campaigner, Derrick Wood, a Customs and Excise official whose mastery of his case, and ability to do battle with other people's, won him frequent commendation from the Commission and from his opponents. His part in the story is even more commendable since attempts were made to 'warn him off'; his Customs and Excise superiors decided that his activities were becoming too political and were no longer compatible with his role as a civil servant. Wood was resigned to leaving the service rather than dropping his campaign until a decision, apparently taken at ministerial level, produced a working compromise.

The series of local inquiries ended at Bedford on 11 September.

('It is with great relief,' said Hunt, 'that I close the last of these local hearings.') Had the Commission been politicians, they would undoubtedly have been tempted at this point to drop their interest in Cublington. There had already been local battles which had stopped governments building reservoirs essential to the need of urban communities but gravely objectionable to rural ones; there had been the Stansted uprising, which, though not decisive alone, had clearly helped to break the original decision. But it was now plain that the force of opposition at Cublington would outweigh anything before it, and – even discounting an element of bluff and rhodomontade – that opposition might not stop short of violence.[22] But the Commission were not politicians. That was the whole point about them. The reports which Hunt had brought back from his pilgrimages were simply one more element to be entered up in the vast equation they were constructing; not to be ignored, but to be rated proportionately along with everything else. Politicians are content with the possible; but Roskill was in search of the perfect.

The results of Roskill's detailed research and investigations were published, to a far from rapturous reception, in the early months of 1970. It was now that the academic onslaught on the cost benefit analysis began to develop, and a process of continuous reappraisal, which lasted almost until the end, was set in hand.

The final series of public hearings began at the Piccadilly Hotel in London on 6 April and ran until August. There were now two parallel battles being fought: one, by the lawyers and the consultants before Mr Justice Roskill in his gilded basement; the other in the country. The latter became at times as rough and bruising as the former was courteous and respectful. By the end, the Commission and the circus of barristers, witnesses and staff had become increasingly a friendly and compact group (though there were gradations even in this; the Foulness protestors, for instance, had a much greater feeling of camaraderie with the protestors of Nuthampstead than they did with the activists of WARA). On 25 June, a cricket match took place between the villagers of threatened Paglesham and a team representing the

Commission, though unhappily not including Sir Eustace him-himself, or for that matter, Colin Buchanan.[23] From August onwards, the Commission was at work on its Report. But on 20 November, hopes of an agreed submission to the Minister were finally destroyed. Colin Buchanan, so long the odd man out, told them he would be unable to put his name to a recommendation for Cublington.

This was not unexpected; indeed, it had always been on the cards, ever since the dispute over the short list at Exeter College two years before. But Buchanan's disillusion went further than had been suspected. Not only would he disavow the recommendation; he would withhold his *imprimatur* from the entire report, with the single exception of the recommendation on the timing of the operation. The majority's dismay was all the greater because Buchanan had been allowed to exert a dominating influence on some sections of the Report, especially that on planning. Yet even this, so nearly his own work, was now to be repudiated. It left a feeling of bitterness which could still be sensed among some of the Commission two years later.

The inevitable result of Buchanan's defection was that the launching of the Roskill Report became as hazardous a proceeding as putting to sea with a time bomb attached to your hull. But the Commission further hampered its chances of success by a decision which, while a tribute to the honesty and openness of its entire operation, nonetheless demonstrated a certain unfortunate naivety. Aware that there had been leaks of its findings before, it decided to release its recommendation without waiting until the full supporting argument was ready for publication.

What appeared on 18 December 1970 was a twenty-one page summary of the Report. Advance copies of the full Report were later sent to the Press and to MPs, but it was not until 21 January that it was finally published. By that time, the press summary had been so exhaustively criticised and condemned that the full Report was mainly of academic interest.

This would have been a dangerous proceeding for any Commission, but with Roskill it was especially so. The whole point of the Roskill findings was that they were built around an intricate balancing of every relevant factor the Commission could lay its hands on; a much too intricate exercise for anyone to summarise in twenty-one pages. The recommendation for Cublington would – for reasons which will appear later – have almost certainly been overthrown anyway; but the method in which the findings were put out ensured that the chances that the Commission's recommendations would succeed were reduced from slim to non-existent.

9

... In which politics has the last word after all

There was never any chance when the Roskill Report finally appeared in January 1971 that the recommendation of Cublington would be accepted. Politically it was already stone dead; killed by a combination of mounting public concern about the environment and the existence in Commons and Lords of a well-marshalled lobby which would have defeated the Government had it tried to go ahead. Much of the money, organisation and full-throated passion which went into defending Cublington between the time of the Roskill finding and the eventual Government decision was directed at an enemy which had already been driven from the field. It had to go on, of course; had there been no bonfires on the beacons, no messages of resolution to fight till the last, no concerted and sustained pressure from local authorities and back-bench members, then Cublington might have begun to looking tempting after all. But the irresistible indications of the evidence are that, by the time the results of the 1970 election were known, it was already very nearly certain that none of the inland sites recommended by Roskill had a chance of acceptance; that by Christmas 1970, even Thurleigh, which was politically the most favourable of the inland sites, was also safe; and that the eventual Cabinet decision was taken with a speed and ease only given to those who are gratefully acknowledging the inevitable.

Roskill was unanimous on the timing of the need for an airport. The arguments of the British Airports Authority that there was extreme urgency to get started were rejected; and in rejecting them, the Commission finally exploded the arguments used by the Government to defend its decision to stand by Stansted in 1967. Everyone on the Commission was agreed that the right time for the first plane to fly off the first new runway was 1980. What they disagreed on, of course, was where that runway should be.

Nuthampstead was clearly out. It was the nearest to London of the sites and the second cheapest to construct. But it was much the worst for noise – indeed, it was only the underestimation of the noise nuisance it would create which had enabled it to get on to the short list. It was the worst on regional planning grounds. It was the least accessible by existing railways and would cause the most serious loss of agricultural land. Douglas Jay's suggestion that Nuthampstead was merely Stansted by another name had blown up in his face, and the Commission themselves underlined it; it was plain, they said, that even if Stansted had reached the short list, it would not have won the nomination.

Next came Foulness; and the Commission's treatment of Foulness rang with the unwritten assertion of their duty to stand firm against a powerful, but often ill-informed, pressure group which had swept too many people along with it in gay oblivion of the more unfortunate facts. They were certainly not blind to its attractions; no encroachment on land, no demolition of property and much of the noise nuisance confined to the sea. Compared to the inland sites, the cost in homes, churches and community life would be small.

But the suggestion that Foulness was also the perfect solution in planning and environmental terms was far too simple. It was wrong anyway to say there would be no immediate loss – the Brent geese were the clearest example of that. But the supposed environmental advantages began to dwindle when you looked at them – and this was the whole philosophy of the Roskill majority – in wider terms. 'The environmental advantages claimed for the choice of Foulness cannot be assessed by considering Foulness in isolation. The repercussions throughout the airport system of this country must also be kept in mind.' The noise nuisance of Foulness, plus that created by continued use of Luton, would outweigh the noise effects of a Cublington or Thurleigh airport which would shut Luton down. And to choose Foulness rather than an inland site would mean that Manchester's Ringway airport, Birmingham's Elmdon, and possibly Gatwick and Stansted too, would also grow. Didn't the rights of householders there need to be considered alongside those of the threatened areas of Buckinghamshire and Bedfordshire? Many of the inland protestors seemed surprisingly insensitive to the effect of the policies they advocated on the ears of people elsewhere: 'many seem to believe that a new inland airport at the edge of the London metropolitan area was an outrage but that a similar development on the outskirts of Manchester or Birmingham would be welcomed by the local populace.'

But the main objection to Foulness was much simpler. It was no use building an airport which met all the necessary tests of environmental acceptability but which then failed as an airport. It was no use locating your airport in a place to which neither airlines nor passengers would want to go. The distance and inaccessibility of Foulness were too much of a deterrent – a point which could be gauged by looking at a map, perhaps, but which had now been spelled out with much greater force by the objective calculations of the cost benefit analysis.

It would cost much more than its rivals to build: £50 million more than Thurleigh; £32 million more than Cublington. Passengers and freight customers would find it inaccessible – so much so that in 1991 some twelve million more air journeys would be made through a Cublington or Thurleigh airport than through Foulness, with the gap widening to thirty million by the year 2000. The extent to which traffic would differ between the possible sites was a crucial indication of the likely success or failure of one site against another. Foulness supporters argued that there was a straightforward way out of this situation; you could simply force the airlines to use Foulness. But they might then respond by using alternative sites out of London, and possibly by not coming to Britain at all; and other governments might retaliate by forcing our own airlines to use inconvenient airports instead of convenient ones.

It is relatively easy [said the Roskill majority] to apply negative controls; but it is unrealistic to attempt to force airlines

13

to operate from a particular airport as long as other alternatives exist and as long as the overseas operations of British airlines are open to unfavourable treatment by foreign governments.

High speed rail links could make the airport more attractive. But if they were cheap enough to interest passengers, a heavy subsidy would be needed; and if there were no subsidy they would be prohibitively expensive. It was said that powers might be given to write off losses, as they were to other nationally essential operations which suffered from chronic states of deficit; but was that really the kind of way in which we should approach such a decision?

Where, as with the third London airport, the nation is about to embark on a huge new investment lasting for many years, not only in an airport but in roads, railways and other services, such arguments must be unacceptable, not least in a country which for so many years has lagged behind others in its economic development. The nation cannot afford to decide the site for this airport on the basis of a serious mis-allocation of scarce resources. There is a substantial risk that Foulness would never produce an adequate return on the substantial capital sums invested in the airport and surface access links and would therefore become a liability to the taxpayer.

Only if the interests of those on the ground were held to exercise an absolute constraint over the interests of those travelling by air could such arguments be ignored; and the Commission did not accept that such constraint existed.

'An airport cannot serve any social purpose unless it first succeeds as an airport.' Foulness could not be certain of success, as Cublington or Thurleigh would be. Economic growth should not be accorded an overwhelming priority; yet if we sacrificed it, we sacrificed too all the benefits it could bring; we would have less money to spend on other necessary priorities – including the environmental ones.

So it lay between Cublington and Thurleigh. Thurleigh would serve the Midlands rather better than Cublington, Cublington was more attractive for people in the South-East. Thurleigh had substantial disadvantages in defence terms and would interfere with scientific establishments. Thurleigh scored on planning grounds, where the problem of Milton Keynes told heavily against Cublington. The disruption of communities (as documented by the Essex University study) and the environmental damage would be worse at Cublington. But Cublington would be better for air travellers (it was supported by the BAA, and by BEA, though BOAC had come out for Thurleigh) and would do more to alleviate the nuisance of Heathrow. It was clearly a very close-run contest; but two factors swung the choice to Cublington. It would make substantially less demand on the nation's resources; and the benefits to air travellers would be greater. Cublington, the Report added in a sentence which had an unusually subjective ring about it 'answers the description of a London airport better than Thurleigh'. The environmental disadvantages were many and serious: 'but we do not believe that they are so serious as to outweigh what we regard as the outstanding advantages of Cublington as a site for a third London airport.'

The Commission then recorded Buchanan's defection and defended itself against his charges. They hoped their Report had demonstrated that they were not the slaves of cost benefit; and they rejected the claim that materialistic considerations had been unduly prominent in their thinking. 'The economic benefits of air travel reflect its contribution to meeting the needs of millions of people, and these needs include a wide spectrum of aesthetic and cultural activities.' As the Commission had noted a few pages before 'we see no reason why the pursuit of leisure should not allow a flight to Rome to see the Sistine Chapel but should permit a car journey to Audley End or Waddesdon Manor.'

The fundamental difference between the majority and Buchanan was that, while they refused to accept any overriding constraint on their decision, he not only accepted but positively gloried in it:

I believe it would be nothing less than an environmental disaster if the airport were to be built at any of the inland sites, but nowhere more serious than at Cublington, where it would lie athwart of the critically important belt of open country between London and Birmingham. [And again:]

I think this case is as critical a test of our sincerity in respect of two related matters as we are likely to be confronted with during this century. The first is our determination to plan comprehensively and with social purpose the commercial and other developments that modern life requires, and the second is our concern for the environment of our not-very-large and densely populated island ... The choice of Cublington would be a grievous blow to conservation policies. It is not merely that there would be a direct setback in the area influenced by the site, even more serious would be the general sense of disillusionment that would come to every person and organisation labouring in the conservation movement, and come just at a point in time when the urgency of the subject becomes daily more apparent. On the other hand, a decision which conceded the importance of the environment (as would be the case if Foulness were chosen, even allowing for the losses involved) would be an event of great significance for the future of Britain. It would show that this country at any rate, in spite of economic difficulties, is prepared to take a stand. I believe such a declaration of attitude would rebound to our credit in more senses than one.[1]

Planners like Abercrombie had taught us the importance of London's 'open background'; it was a snug, homely and liveable part of the world, full of treasures and of a deep quietude. For decades, successive governments had sought to conserve it. It had absorbed new towns; but to try to fit into it a four-runway airport, covering an area as great as that bounded by Paddington, Victoria, Liverpool Street and King's Cross, was a demand upon it which was without precedent.

Buchanan also defended his preference as carrying a clear

benefit in regional planning. To do this, he had to confront a difficulty. Evidence on the regional planning implications given to the inquiry by Dr Wilfred Burns, the Ministry's chief planner and head of the South-East Joint Planning Team, had made little distinction between the merits and demerits of the four sites.

The treatment of this issue was, like the absence of a national airports plan, a large unwelcome complication for the Roskill Commission. Regional planning had, since the mid-sixties when no government expert on it was called before the Blake inquiry, become almost a vogue; but the vogue had come a little too late to help Roskill. The Joint Planning Team was still at an early stage in its work, and there were great fears about any evidence being given by it at all. The whole exercise had been devised as a result of sensitive negotiations between Whitehall, the Standing Conference on London and South-East Regional Planning (itself representing all the planning authorities of the region) and the South-East Economic Planning Council. No findings had yet been reported back to anyone. For the leader of the team to get up before Roskill and draw all sorts of conclusions from what it had learned so far might have dangerous implications for future policy and might infuriate some of those whose consent to the exercise the Ministry had spent so much time in carefully coaxing.

The difficulty was considered so serious that the Commission decided to consult Crosland about it. He suggested that the Commission's operation might have to be slowed down to give this evidence time to materialise, but not surprisingly the Commission felt they were giving enough of their lives to the inquiry already without accepting that. On the other hand, they made it clear that they did not feel they could do their work properly if they did not hear from the team at all. So a compromise was reached. Burns was allowed to testify, but he would have to be discreet, and he would have to give his evidence 'in a personal capacity'. The result was evidence which, for all the known expertise of its author, was somewhat tentative. Nuthampstead was worst; but there was no overriding reason for preferring

Foulness on regional planning grounds to Cublington or
Thurleigh.

There had been during the inquiry some signs of conflict
between the Commission's own expert in this field, Buchanan,
and the outside expert whom they had been at such great pains
to parade before them. And although Buchanan was allowed to
exert a formative influence on the chapter on planning, the
majority found themselves in the end defending Burns against
him: 'We believe it was right to follow Dr Burns's evidence and
unlike Professor Buchanan we would have felt gravely handi-
capped had it not been made available to us.' Buchanan, they
noted, perhaps a little bitterly, had appeared to be united with
them in their approaches to Crosland at the end of 1969 and early
in 1970 to get Burns's work hurried on so that they might draw
upon it; but now he was ready to prefer his own predictions and
assessments to it.

Where Burns's guidance was, understandably, careful and
often negative, Buchanan's conclusions were positive and bold.
Foulness, he argued, could bring to the depressed region north of
the Thames Estuary – that 'huge, obdurate mass,' as he called it,
'of depressed and impoverished development' – the economic
stimulus it had lacked for so long. Burns was worried about
cramming so much in to the confined space of the south Essex
corridor and was particularly alarmed by the possibility of
development spreading across the River Crouch. Buchanan's
own assessment 'as a professional planner' was that with careful
planning the thing could be done.

With equal boldness, Buchanan set about the argument that
Foulness might be a bad risk, a potential white elephant, and
replied that if this was so, it would only be because of a failure
of nerve. He did not believe a single tourist would be dissuaded
from coming to Britain because of the remoteness of the site;
and if that implied disbelief of the cost benefit analysis, then this
was a charge to which he would readily plead guilty.[2] It had
been pushed too far and too fast beyond the fairly easily quantified
problems in which it usually operated.

From this base, he was able to launch another crucial assault; on Roskill's concept of balance. The majority had made much of consequential effects on Luton: 'I think we are all agreed that the location of the third London airport could not possibly be allowed to be dictated by considerations arising from a small municipal airport.'[3] He admitted it might continue; but it could be subjected to strict noise controls. Cublington, it was said, could relieve Heathrow:

> but I can see no reason why, if Foulness were chosen, it should not be possible by treating the airlines with great firmness to make it perform as well as Cublington in this particular respect . . . I do not accept [Buchanan concluded] that Foulness would be economically unsound provided it is seen in the context of a sensible national airports policy firmly administered.

And so, six months after its arrival in office, the Conservative Government was confronted with a situation in some ways reminiscent of that which faced Labour with the report of the Donovan Commission. In each case, the Government had turned to an outside body in search of a consensus basis for an important political decision. In each case, the result had been a majority report of great authority, joined with a minority view which was also deeply persuasive. In each case, the majority view was based on detailed and intricate argument; in each case, it relied heavily on a research effort which was subjected to much public criticism.[4] And with Roskill as with Donovan, the dissenting report – Andrew Shonfield's 'note of reservation' and Buchanan's 'note of dissent' – chimed better than the majority's with an accumulating political fashion. Shonfield, unlike his colleagues, shared the taste (shown by the opinion polls to have developed among people at large, and even among trade unionists) for greater state regulation of the way unions behaved. Buchanan came nearer than Roskill to catching the characteristic political mood of the end of the sixties; a feeling that growth had been pursued too far and that the claims of the environment had been too much neglected.

In each case (as also with Derek Senior's dissent from the Redcliffe-Maud Commission) the emergence of one influential member out of tune with his colleagues acted both as a focus for public attention and as a base which the Government could occupy in overthrowing the conclusions of the main report.

Buchanan provided the philosophical and intellectual justifications for the forces which united against the Roskill recommendation. These forces were three: the new Department of Environment, which under its ambitious young Minister, Peter Walker, shared Buchanan's belief that this was a test for whether we in Britain took the environmental issues seriously; the strong environmental lobby which had developed in the country, together with the smaller but especially passionate army in north Buckinghamshire which had invoked their aid in the defence of their homes and their way of life; and the Parliamentary forces which had determined to defeat the Roskill recommendation.

Walker was later to claim that the very existence of the DOE, uniting as it did environmental and transportation departments under one roof, made a right decision on Roskill more likely than it would otherwise have been. If this was so, it was not in the sense that conflicts which might have occurred between departments over the Roskill recommendation were obliterated by the creation in October 1970 of the department over which Walker presided. That certainly happened in some cases; the old Ministry of Transport would probably have been much more permissive, for instance, about heavy lorries than the new DOE was now.[5] But the conflict here was between the economic interest, represented by the old Board of Trade, now incorporated in the Department of Trade and Industry, and the Department of the Environment, not between different sectors of the Walker empire. The real significance of the changes was that Walker fought from a much more powerful political and governmental base than those who had looked after planning interests before.

What was more, he was a much shrewder and more experienced political operator than his opposite number in Trade and Industry. Had it not been for the death of Iain Macleod, the cast list for this encounter would have been rather different. But Macleod's death meant the removal of Anthony Barber from Europe to take the Treasury, and the transfer of Geoffrey Rippon from Trade to Europe. Into Trade came the former Director-General of the CBI, John Davies, who had not entered Parliament until the 1970 general election. Though no one who had led the CBI could be a stranger to political infighting and the political framework, he was still, as he then regularly demonstrated, somewhat out of his depth.

The sudden advance of the environment as a political issue – which was both signalled and further impelled by the creation of the DOE – also strengthened Walker. It had been stirring for a long time now. Its growth is difficult to trace, but there are certain obvious landmarks; the Stansted story itself, and before it the saving of Ullswater in 1962, had each demonstrated that the environmental issue was growing bigger than governments appeared to have bargained for.

The new environmental activists were a coalition of different preoccupations and different kinds of emphasis. We had, they said, sacrificed too much of our environment to mindless development. We had seen the character of cities like York and Worcester damaged if not destroyed by the vandalistic pulling down of valuable old buildings and the construction of new ones which looked as though they had been taken off a shelf in the developer's warehouse, which bore no relation to the surroundings into which they fitted, and which were slotted in without distinction into streets of the most sharply differing character. Industry, they argued, had grown too heedless of its responsibilities and must be made to curb the pollution it caused. They were, by and large, in favour of deliberate state policy to reduce the birth rate.

Some of them went further; they asserted that growth was a false god which had been whored after for too long; and that if we cared for our environment we must be ready to sacrifice the greater prosperity which was the most widely advertised reward of the expanding economy. The grim and despairing prophet of this view was E. J. Mishan, whose book, *The Costs of Economic*

Growth[6] seemed at first to be an extreme statement of the case, though it was one which began to look less extreme as time went on. The faith which had been proclaimed by a long line of airport champions, from Mrs Tate and her friends in the forties, through the authors of the White Papers on Gatwick and Stansted in the fifties and sixties, down to the Roskill Commission itself, determined as any that we should not shed our status as the Clapham Junction of the air, was a natural target for these campaigners.

Much earlier, when this line of thinking was still relatively unfashionable, Lord Plowden, who as chairman of the Economic Planning Board from 1947 to 1953, and as chairman of Tube Investments, had sought growth in his day as avidly as any, challenged the assumption that the country ought necessarily to welcome unfettered aviation. The fundamental question, he had suggested in the 1967 Lords debate (with all the added fervour of one whose home was at Great Dunmow, well within the area which Stansted threatened to blight) was what price the community ought to be ready to pay for technological progress. Once, we had sent women and children down the mines in the name of growth. Today, we assumed that cars should be made as cheaply as possible and used by people where and when they pleased, 'but we are still struggling with the social cost this is imposing on us'. He went on to propound a rare heresy. We actually had the choice, he said, of saying no to air traffic: 'it is not self-evident that it would be a material disaster if some air traffic were diverted to the Continent, or even to the Midlands or the north of this country.' In 1967, hardly any politician could be heard taking as sharp a line as that; by 1973, it was possible to state it publicly without fear of being declared eccentric.[7]

In the late sixties concern with the claims of the environment began to take up a steadily increasing ration of the time and attention of Government. That great environmental epic, the wreck of the *Torrey Canyon* in March 1967, publicised the dangers of pollution as no amount of campaigning by the environmental activists could have done. When Labour set up its new Depart-

ment of Local Government and Regional Planning under Crosland in October 1969, Harold Wilson announced that one of the
main duties of the new department would be to look urgently
at the question of pollution in all its forms. In December 1969, a
standing Royal Commission on Environmental Pollution was
set up under Sir Eric Ashby, Master of Clare College, Cambridge,
though its initial findings were too mild for the taste of some of
those who had called for its establishment.

The crusade had by now begun to attract the newspapers, which
took to lecturing the politicians on the need to pay serious attention to the questions of preservation, pollution and population.
In February 1970 the BBC launched a television series called
Doomwatch in which, week by week, viewers were exposed to
various brands of potential ecological disaster. The heroes of
the programme were a group of scientists who were forever
rushing off at a moment's notice to investigate some form of
environmental villainy and see that it was stopped. This programme helped to develop a taste for knights in shining armour
who could ride in and deal swiftly and, if necessary, autocrati-
cally with any threat to the environment. Organisations like the
Alkali Inspectorate, which took pride in proceeding by persua-
sion rather than by prosecution, found their records of action in
the courts, paraded about disparagingly like the League perfor-
mances of unsuccessful football teams. Their fuddy-duddy kinds
of compromise would not do in the new ecologically-minded
world of the end of the sixties.

The movement was perhaps most tellingly epitomised by the
success of a magazine called *The Ecologist* which appeared from
nowhere in 1970 to achieve an estimated sale at one point of
17,000 a month. Its *Blueprint for Survival* – a blueprint based on
breaking with the pursuit of growth – commanded hundreds of
column inches early in 1972 and brought its authors a summons
to the presence of Peter Walker. The movement brought new life
to the Conservation Society, whose membership grew from
1,300 in 1969 to 5,700 in 1971. At one point, new environmental
pressure groups appeared to be springing up almost every week.

In January 1972, the *Daily Telegraph* reported plans to set up an Ecological Foundation with leading scientists and businessmen as trustees; an Association of Environmental Teachers had also been founded with 1,300 members but no funds.

It would be wrong to suggest that the environment had become a dominant political issue by the time of the 1970 general election. There were not, one suspects, many votes swung by it, particularly since it was difficult anyway to be certain that one party was that much more environmentally-minded than another. Yet politicians returning after the election confessed themselves impressed by the way these questions kept cropping up at meetings and on the doorstep; and they became aware that to seem indifferent to the claims of the environment might now be to incur an unwelcome electoral liability – particularly among the young, who, being the most pragmatic element in the whole electorate, and having the longest voting life ahead of them, had a significance in electoral calculations out of proportion to their numbers.

The outcry provoked when Peter Walker – then still a mere Minister of Housing and Local Government (the Department of Environment was created in October 1970) – approved a plan for a brewery at Samlesbury in the countryside between Blackburn and Preston was a telling indication of what was happening. Compared with the decision on the third London airport, this was a matter of quite monumental insignificance. But it was taken as a symbolic statement of the intentions of the new government, a pointer to the way in which future decisions might be expected to go; and Walker found himself attacked for his supine failure to defend the environment, whose protector he was supposed to be, from the pressures of commercial interests (interests which, it was sardonically observed, had contributed not ungenerously to the funds of the Conservative Party).

Buchanan's declaration that a government which really cared about the environment would be bound to reject the decision of the Roskill majority was therefore one to which the Government in general, and Walker in particular, were particularly sensitive. And the growth of environmental consciousness ensured that the campaign of the people of the Cublington area would fall upon thousands of receptive ears, commanding the sympathy of far more people than the earlier campaign for Stansted could have done. It had been possible, though unwise, to shut one's ears to the pressure over Stansted; but no one with any instinct for politics, or even for survival, could have ignored the strength and depth of the clamour over Cublington.

Foulness, of course, involved environmental losses too. Buchanan himself had recognised that. But those who might have been expected to proclaim the dangers at Foulness were already enlisted on platforms elsewhere. One by one, the influential environmental organisations had lined up in the 'inland' camp. Like Essex County Council, they had settled on Foulness as the sacrifice which would preserve them from losing places still more precious to them. Bodies like the Nature Conservancy, the Civic Trust and the Council for the Preservation of Rural England, to which people defending their rural heritage would normally have turned, could offer no help to Derrick Wood and his comrades in south-east Essex. The National Farmers Union, defenders of agricultural land, were committed to the view (even though an eminent consultant hired by Roskill at their behest came to a different conclusion[8]) that Foulness, regrettable though its loss would be, was the best available choice.

Only the wild life groups, especially the Royal Society for the Protection of Birds, lined up on the side of the defenders of Foulness; the sacrifice of the bird life there was, to people like the naturalist Peter Scott who testified as passionately against it as John Betjeman had testified against Cublington, an environmental disaster of the same proportions as Buchanan had foreseen at the inland sites. But against the array of interests on the other side, they carried little weight.

The newspapers were unanimous in condemning the choice of Cublington. The majority of their commentators backed Foulness, though there were already some who declared that a third London airport was no longer necessary. The defenders of

Foulness found much of what was said and written in the news-papers hard to take. What infuriated them most of all was a BBC broadcast which claimed to feature the voice of protest at each of the four sites. When they came to Foulness, they played a recording of gulls. Foulness was frequently portrayed as though it was a painless solution. A classic example of this tendency occurred in the *Daily Telegraph* (whose lavish coverage of the Cublington protestors had also infuriated the Foulness people) on 21 December:

> The choice, according to one noise expert, is simple. Either Britain seized the opportunity at Foulness of leading the world in commercial and inventive skills, or risked a nation of idiots, half crazed with the ever-increasing noise of planes passing overhead.

True, destruction on site would be far less than at Cublington. But as one of the few public defenders of Roskill, Christopher Foster, head of the Centre for Urban Economics at LSE, who had been Director-General of Economic Planning at the Ministry of Transport from 1966 to 1970, wrote in *The Times* on 4 March 1971, when you added in the building of link roads and the creation of an airport city it began to look rather different.

Even in the name of fair play, there was – the defenders of Foulness felt – discrimination against them. Radio and television broadcasts after the Roskill short list was published habitually consisted of one representative from each of the short list sites. Equally consistently, the three inland representatives would un-animously agree on the necessity of building at Foulness, leaving only one Foulness militant to oppose it.

The build-up of environmentalist pressure against Cublington and for Foulness had therefore made it impossible for the Govern-ment to accept the findings of the Roskill Report without alienat-ing, perhaps in some cases permanently, a lobby which, both in its grass roots support and in the involvement in it of influen-tial institutions, was gaining in political significance all the time.

But the direct political confrontation involved in choosing Cublington was, if anything, even more daunting. Here again, the choice of three inland sites against one estuarial site in the Roskill short list had immensely strengthened the likelihood that Foulness would be the eventual choice. The 'popular' airport groups like WARA and BARA had seen from the start the advantages of forming a single united front and declaring, in effect: 'Hands off us, and hands off them as well.' Under the concentrated scrutiny of Roskill, there was an occasional wavering; pressed by the Commission to say which of the inland sites would be the best choice, counsel for WARA, Niall MacDermot, reluctantly but very definitely commended Thurleigh. But that kind of disloyalty only occurred under pressure.

In the same way, the county councils which opposed each inland site had announced their equal opposition to others, coupled with enthusiastic advocacy of Foulness. Choosing an inland site thus involved earning the deep hostility of at least half a dozen major local authorities, not to mention dozens of smaller borough and district authorities. Choosing Foulness would involve offending none.

Even more crucially, the MPs involved in the fight to save Cublington, Thurleigh and Nuthampstead had joined forces in a single organisation, known as the Inland Group, which implacably signalled to the Government its resistance to the choice of any inland site, and took care to keep the Whips' office posted about its readiness to vote down the appropriate order if that day ever came.

There had, of course, been a similar 'revolt' over Stansted in Labour's day and it had come to nothing. So on the face of it, one might have expected the Whips to be sceptical now. But the Inland Group, chaired by Stephen Hastings, the astute Conservative MP for Mid-Bedfordshire, who had constituents threatened by all three sites, and working closely and productively with leaders of WARA, was a much more formidable political force than the impromptu alliance of MPs which had signed the early day motion on Stansted.

If there had ever been any doubt about that, the Conservative gains in the 1970 general election removed it. Only a handful of Labour back-benchers had any direct constituency interest in the Stansted proposals – and in face of a Labour majority approaching a hundred, they were understandably regarded as of negligible importance. But Cublington, Thurleigh and Nuthampstead – all these were set in areas of predominantly Conservative representation; and with the Government's majority only thirty-one they were too formidable to be ignored.

It is worth setting out the figures, because they clearly demonstrate that Roskill's recommendation had no chance of succeeding. The prospect of a new airport seems to cause the greatest concern in constituencies within a fifteen-mile radius of the chosen site. There were eleven constituencies within this radius of Stansted in 1967 and only three were Labour held.

There were fifteen constituencies with fifteen miles of Cublington in 1971 of which thirteen were represented by members of the government party. But because of the formation of the inland alliance, one could add the MPs threatened by Nuthampstead and Thurleigh to the defenders of Cublington. That gave a total of twenty-five constituencies with twenty-two Conservative MPs. Not all could be relied on to vote against the Government, of course; the Government would have been pretty sure, for instance, of the support of the MP for Cambridgeshire, Francis Pym, since he happened to be the Chief Whip. But no government could cheerfully have done battle on such a contentious issue with a group as strong as this.

However, the Inland Group had taken the extra precaution of recruiting a number of MPs who were active campaigners against existing airports. This was a particularly shrewd move in that, on Roskill's calculations, these people should have been vigorously backing his recommendation; he had, after all, argued that to sacrifice Cublington would bring relief to people around Heathrow, Gatwick and Luton who would get no such respite from the choice of Foulness. But this was not what happened. Conservative back-benchers like Toby Jessel (Twickenham),

Carol Mather (Esher) and Sir George Sinclair (Dorking) who represented constituencies where there was constant complaint about Heathrow and Gatwick, were to be found in the Inland camp.

The 134 Conservative MPs who signed the Inland Group motion therefore constituted a much more viable set of parliamentary shock troops than the authors of the Stansted motion. And it took no great statistical expertise to work out how many thousands of potentially affronted Conservative votes the inland MPs and their Heathrow and Gatwick comrades must represent. The motion declared 'that this House, while recognising the need for a third London airport, is totally opposed to the choice of any inland site, or to the extension of any existing airport for this purpose; and strongly advocates the selection of Foulness or any other suitable site'. It was tabled on 18 December, the day the Cublington recommendation was announced (they did not wait to see the full Report) though preparations had been laid long before. By Christmas there were 169 signatures on it and the final total reached 219, with one MP whose constituents were overflown from Heathrow getting his name on twice.

There could be no matching counter-demonstration for Foulness. There were thirteen constituencies within fifteen miles of the Maplin Sands site, seven in Essex and six in Kent. Twelve had Conservative MPs – so there was the making there of a useful team of rebels, though not on the scale of the Inland Group. However, the MPs most closely concerned had not taken a militant anti-Foulness line. Bernard (later Sir Bernard) Braine (South-East Essex) had raised many objections in evidence to Roskill but had never opposed the project out of hand. He and Sir Stephen McAdden (Southend East) concentrated on trying to get the airport site pushed as far north-eastwards up the Maplin Sands as possible – and they were rewarded when the Government eventually designated a site well to the north-east of that plotted by Roskill. The MP for Southend West, Paul Channon, was a Minister and could not campaign at all. In any case, the MPs in this area, unlike those in north Bucks, knew that many of

14

their constituents would not welcome outright opposition to the airport.

Another Conservative MP on this list, Norman St John-Stevas (Chelmsford) had actually signed the Inland Group motion; more of his constituents were likely to be opposed to Nuthampstead than were likely to object to Foulness. The strongest opposition to Foulness came from the Kent MPs, whose constituents had much to lose and nothing to gain from a Foulness airport – notably David Crouch (Canterbury), Roger Moate (Faversham) and Roger White (Gravesend). But this useful nucleus failed to attract the much wider support which the Inland Group commanded. A motion supporting Roskill's recommendations was duly tabled on 26 January – after the full Report appeared – by a group including Crouch, McAdden, White and Brian Harrison (Maldon) but its handful of signatures only served to emphasise the unbeatable lead which the Inland Group had taken.

The same physical circumstances which made Foulness so popular as an airport – its station on the very edge of England – made it harder, politically, to defend. Its fifteen-mile radius included a large area whose only constituents were geese. 'I regard the countryside as a natural heritage,' said Toby Jessel, Conservative MP for Twickenham, in the Commons debate of 4 March 1971, 'and there is not so much left unspoiled that we can afford to see sizeable chunks of it destroyed.' Yet the coastline is part of our heritage too; and an even smaller proportion of that, in south-east England, has escaped destruction. Anyone who knows the north bank of the estuary around Tilbury is familiar with the devastation which industry can inflict upon it. But there are few votes in the coastline; there are no predominantly coastal constituencies, as there are rural constituencies; the politics of the coastline are a minority affair, which makes it perpetually vulnerable.

Even had the line-up in the Commons not been so formidable a deterrent (and there is no doubt that the Government's business managers were certain almost at the moment that the Roskill recommendation appeared that there was no chance of getting the Commons to approve it) the case would have foundered in the Lords, where a stalwart group who had been advocating Foulness for nearly four years now had no intention of giving up at a time like this. A debate in the Lords on 22 February produced the same almost unanimous procession of condemnation for Cublington that there had once been for Stansted.

Many minds had been made up long before. Lord Reigate reported that he had recently met Lord Roskill quite by chance and told him that he was in favour of Foulness. 'I hope that you have read the Report and I hope all who speak will have read the Report,' Lord Roskill had replied. 'At that time, I had not read it,' declared Lord Reigate. 'I have now and I am still in favour.' Verdict first, consideration of the facts later on; it was exactly what the defenders of Foulness suspected. In the House of Lords, as in the outside world, there seemed to be few who had read and digested the whole of what the Roskill majority had said. A lot of people seemed to have read Buchanan; but that, of course, was shorter and less intellectually demanding than the majority Report.

Had the Government been united behind the Cublington proposal it could, even now, have weighed up the chances of trying to bulldoze the Roskill recommendation through. But that wasn't the case either. Officially the Government remained neutral throughout, simply asking Lords and Commons to take note of what the Report had said. Peter Walker made a speech in the Commons debate of 4 March, which looked at first sight to be impartial. Yet both here and in interviews, Walker attached a weight to the environmental aspect of the decision which would have left him in an embarrassing position had Cublington been chosen.

Meanwhile, the arrival of the bad news at Cublington had been greeted with the explosions of wrath which anyone who attended the public hearings might have expected. Well to the fore was the Reverend Sillitoe of Dunton, who gave a characteristically warlike interview to John Ezard of the *Guardian* on 5 January:

> I would give my blessing to people who fought because I believe it is licit for a Christian to bear arms in defence of his home ... If it comes to the final issue, which God forbid, then, as in 1940, we will fight on the doorsteps of our homes, in the fields and the farms, at churchyard gates and at church doors. Come the four corners of bureaucracy and we shall shock them ... I would say as a priest that although people who fight might be legally wrong they would be morally right. I was a chaplain in the trenches in the First World War ... I am not a pacifist.

Clarion calls like this, coupled with reports of guerrilla warfare being plotted by extremists, worried the great majority of protestors and caused particular concern to WARA who realised they would be counter-productive. Less ferocious demonstration made the point just as well. One night in January, bonfires burned in villages and hilltops over an area of twenty square miles. Three hundred vehicles took part in a motorcade round the twenty-eight miles of the airport perimeter; eight, ten, even twenty thousand people were reported to have attended a protest rally at Wing. Local authorities held crisis meetings; early in February, Oxfordshire, Buckinghamshire, Bedfordshire and Northamptonshire told the Government that if Cublington were selected they intended to refuse planning permission.

Faced with a plan by a militant minority to shut the whole area completely to outsiders, WARA riposted by declaring an Open Day in which people from all over Britain were invited to visit the area, meet its people, and sample its homely delights. The whole enterprise was a huge success and the crop of column inches which were later stuck into scrapbooks was deeply rewarding.

But in reality the battle was by this time already over. The correct conclusion had been drawn by Evelyn de Rothschild and other backers of WARA who decided, a month after Roskill

reported, that it was no use going ahead with a planned advertis-
ing campaign because Cublington was already safe.[9] In Cabinet,
Walker was never in danger of losing. In the Department of
Trade and Industry there were those who were already reconciled
to Foulness; while others, knowing Cublington was impossible,
were vainly trying to build up a campaign to go to Thurleigh.

Had the Roskill Commission been politicians, of course,
Thurleigh would have been chosen in the first place. The balance
of advantage between Cublington and Thurleigh was slight.
Thurleigh, as the Essex University study had shown, had a less
compact and homogeneous community; it had also produced a
pro-airport movement. It was really a small sacrifice to abandon
the best Roskill site for the second best – much less of a sacrifice
than abandoning it for a site which had rated thirteenth when they
came to draw up the short list.

Even so, the inland alliances, at grass roots, local authority
and parliamentary level, could have been relied on to defend
Thurleigh, and to turn out in greater numbers than the opponents
of Foulness. And since Foulness was there, and since it was agreed,
it seemed, by almost everyone who lived outside the area (and
even some who lived within it) that a Foulness airport was not
just possible but actually desirable, one can be fairly certain that
Thurleigh too would have escaped.

But if Thurleigh had its small band of advocates, there was
hardly a voice raised in favour of building at Cublington. The
Commission were silent. Roskill himself, as a judge, was not in
the habit of going on television, or writing crusading pieces in the
Sunday Express, to defend those of his judgments which had
proved unpopular, and he did not mean to do so now. Nor were
his colleagues to do so; he enjoined on them the same vow of
silence he imposed upon himself. Only rarely was there a word
said for Cublington at all. Christopher Foster, in *The Times* on
4 March, was forthright: 'The report should have been hailed as
a triumph of reason and humanity which might help other nations
struggling with contentious planning problems. It is far in advance
of anything anyone else has done.' But he was arguing a lost cause.

All it needed now for the last piece to fall into place was a willingness on the part of the Government to force traffic into Foulness, either by direction, or by making the alternatives less attractive. And that, it emerged, they were prepared to do – to the extent of baulking the BAA of its second runway at Gatwick and denying further expansion at Luton and Stansted.[10]

The announcement of the decision to overrule Roskill and build at Foulness, on 26 April 1971, was made by John Davies, but the will behind it was Walker's. The Government, said Davies, accepted the need for a third airport by 1980. To defer it on the basis of such technical developments as short and vertical take-off, which were still in a 'speculative' stage, would involve an unacceptable degree of risk.

The Commission had emphasised that Foulness was the best site on environmental grounds.

> The Government have concluded [Davies said] that these considerations are of paramount importance. In the Government's view, the irreversible damage that would be done to large tracts of countryside and to many settled communities by the creation of an airport at any of the three inland sites studied by the Commission is so great that it is worth paying the price involved in selecting Foulness.

There would be economic disadvantages, certainly; 'but the Government are confident that an airport at Foulness will meet the needs of aviation in spite of the economic penalty involved'. The economic disadvantages could be reduced by the building of a rapid transit link with the airport and by efficient operation; and use of the new airport could be 'encouraged' by stricter limits on movements at other airports, which, in turn, would reduce their noise. 'It will be open to the British Airports Authority,' Davies added, 'to arrange charges between its airports so as to stimulate traffic at Foulness.'

Plans would be made on a basis that would not rule out an associated seaport – though proposals for involving private capital were still being looked at. And Davies ended with a promise which – if it were fulfilled – would meet one of the most ancient demands of Whitehall's critics: 'these decisions, dealing as they do with the large part of our air traffic, will provide a basis on which studies can be pursued to establish the desirable pattern of airport development in the rest of the country.'

Walker told a press conference later that the additional cost of Foulness was now put at £150 million – but this was a price that government and nation must be prepared to pay to avoid the disruption of the countryside involved in Cublington and Thurleigh.

The reaction – allowing for a few doubts about the cost – was generally warm, mounting in places to rapturous. The decision, it was said, especially by those who had themselves publicly advocated it, showed that Britain was growing up, learning to look beyond its economic nose and to discern the wider horizon. Few feared that this forward looking posture might lead British aviation, like Thales, to fall yet again into a well. The airlines were displeased, of course, but they soon fell to contemplating the splendid new airport they could create and began to look a little more cheerful. Peter Masefield's appointment as chairman of BAA was not renewed, which removed one influential doubter from the scene, and his successor lauded the new Maplin airport in the next annual report with the same ripe confidence with which Masefield had formerly attacked it. For the first time in nearly thirty years, a major decision in airport policy commanded general enthusiasm. That the politicians were judged to have chosen right, while the outsiders called in to rescue them from the consequences of their earlier mismanagement were all but universally agreed to have chosen wrong, was an irony almost overlooked in the rejoicing.

10

. . . *In which Peter Kirk asks a sensible question, but does not get an answer*

And that might have been that; but, in the way of airports policy, it wasn't. Already before the Government announced its decision there were those who declared that it was now hopelessly out of date. It would, they said, be a white elephant; for anyone who studied the latest technical evidence and set aside all preconceptions would see that the airport which was now planned to be put on the Maplin Sands was not now needed at all.

Planes were getting bigger; and we had made the mistake of underestimating the switch to larger planes before. Latest predictions suggested that the noise nuisance of planes could be cut further and faster than anyone had expected at the time Roskill was sitting. That meant that continuing with, even expanding, Heathrow, Gatwick and Luton might be less unpopular than Roskill had supposed. The new generation of planes – 747s, DC 10s, Tristars – would be much gentler and more agreeable neighbours than their predecessors. (That was one reason for saving Rolls-Royce, whose operations were so critical to the breakthrough to quieter engines.)

Then there were the new techniques in aircraft design. Roskill – one of whose members, Professor Keith-Lucas, was a leading authority in the field – had seen no reason to expect a switch into vertical take-off in time to alter calculations which had suggested that the third airport would be needed in 1980. But now new evidence was accumulating of techniques which advanced on VTOL and also enabled you to operate with much less noise. One had to assume that the change from conventional to reduced take-off flying would not be impeded by the natural desire of the airlines to get some use out of a new generation of conventional planes before consigning them to the scrap heap. But if you were arguing about wasted investment, what about the waste involved in constructing a quite unnecessary airport on Maplin Sands?

Techniques of air traffic control, too, might improve very soon to a stage where more traffic could be crammed into existing space. The most striking figure to take the stand against Foulness was Anthony Crosland. He had attacked the decision as soon as Davies announced it.

> I believe the Government are, on balance, right to have rejected the three Roskill inland sites. [He said] I think they are also right to have rejected a second runway at Gatwick. But the choice of Foulness is totally wrong, on the ground of damage to the environment, and particularly the coastline; on the ground of destruction of homes for motorways; on the ground of enormous additional cost; and probably also on the grounds of safety ... I prophesy that Foulness if it is ever built will turn out to be the white elephant of the century, because airlines will not use it.

He developed the theme more fully in the debate of 4 March 1971. Having made his case that the Commission had tragically misread its terms of reference[1] he went on to argue a strategy for doing without Foulness. This would involve maximising use of existing airports; but he was not ready to subscribe to the argument that you need build nothing at all. A new two-runway airport would be justified. There was no reason, though, to put it at Foulness. A lot of places were more suitable; one possibility would be Cublington. To think again now could save us the cost of a Maplin[2] white elephant, and could avoid destruction which, when the toll of the motorways was counted in, would be worse than Cublington's.

A similar case had been argued in the Lords by two of the House's aviation specialists – Lord Beswick and Lord King's Norton; and the 'do-without-it' school began to declare themselves in increasing profusion in the Press. In January 1973, the debate suddenly burst on to the front pages with reports that Conservative back-bench doubts about the project had reached such a pitch that the Government might even be defeated on the second reading of the Maplin Development Bill.

The trouble started at a meeting of the back-bench Aviation Committee on 24 January, at which the new Environment Minister, Geoffrey Rippon, was thought by many present to have taken too carefree and casual an attitude to the criticisms and complaints laid before him. A report next day in *The Times* said that twenty-four of twenty-seven members who spoke had opposed the project, and suggested that the Government might be in trouble with its Bill. When next day it was discovered that the second reading debate had been removed from the programme of the next week's business, the rest of the Press took up the story.

In fact the story in *The Times* had a strong ring of melodrama about it. The numbers stated were exaggerated, and in any case, not all who criticised could be expected to deny the Government their votes. But in the way that newspaper stories often do, this one took on a quality of posthumous truth. Once the word got about that there might be a serious revolt, MPs began to think more seriously about their own attitudes to the project. *The Times* had a deeply troubled leader about it, and there was a succession of powerful letters to the editor, notably one from the Roskill Commission's economist member, Alan Walters, which added to a string of other condemnations of the plan the suggestion that it might be a boost to inflation – a warning that a government now up to its neck in the battle with inflation could not entirely ignore.

In the course of this correspondence one was able to discern the posthumous triumph of the Roskill Commission. Much of what was said now had been said already in the Report; but the arguments which had been ignored in the placid pages of a document which, in any case, had gone largely unread took on a new glamour when compressed with such bitterness and urgency into the length of a letter to *The Times*.

The objectors within the Conservative Party fell into three main groups. There were the local Members from Essex and Kent, like Braine, Crouch and McAdden; there were those who were impressed by technical arguments, whether about new kinds of aircraft or, like Robin Maxwell-Hyslop, the Member for Tiverton, with the dangers of bird strike; and there were those who wanted to boost regional airports, like Robert Adley, the Member for Bristol North-East and a persistent advocate of an airport for Severnside.

Meanwhile, at first almost unremarked, something significant had happened on the Labour side. The Opposition officially committed itself against the Maplin project.

This happened in a curiously irrational way. A senior Labour politician was asked to say how the Party would vote in the division on the Maplin Development Bill. He said they would be against it, on the lines Tony Crosland had taken. The matter had not in fact been considered either by the Parliamentary Party or the Shadow Cabinet. In the Parliamentary Party, indeed, there were members like Hugh Jenkins and Michael Barnes, with a strong constituency interest on cutting down noise at Heathrow, who had been critical of Crosland's readiness to do without Maplin. The first they knew about their Party's official opposition was when they read about it in the newspapers on 26 January.

The Party's position was then considered at a meeting of the Environment Group of back-benchers, with Crosland in the chair. Here, despite Jenkins's complaints, the Crosland line was over-whelmingly approved. That was reported to the following meeting of the Shadow Cabinet, which then put forward its own recom-mendations. The Party should table a 'reasoned amendment', not totally opposing the project, but suggesting that a halt should be called until a review being carried out within the Civil Aviation Authority had been completed. However, if that reasoned amend-ment were rejected, they should then vote against the Govern-ment on second reading. It was this last recommendation which finally moved the Labour Party into the anti-Maplin camp.

The reasoned amendment was in itself a somewhat illogical

response to the Bill, since the airport, which the CAA was reviewing, was only one element in the project the Commons were being asked to approve. The CAA could hardly be expected to say very much about the seaport, which, since it was now expected to be the biggest of its kind in Europe, was a not inconsiderable feature in the decision which Parliament had to take.

Indeed, this had long since ceased to be simply a question of providing a third airport for the capital, in the sequence which had begun with Stansted and continued with Roskill. The cost, size and scope had now expanded far beyond anything that had been imagined at the time of the Roskill Commission. That was one reason why so many MPs on both sides were so unhappy about it. The cost of the project – about which the Government had for some time been refusing to come clean, despite a series of parliamentary questions about it – was now said to be £825 million, more than our share in the Concorde project; although opponents of the scheme said this estimate was just as inadequate and unrealistic as the early estimates of Concorde had been. 'I do not know how on earth we shall justify to our colleagues in the EEC that they should pour money into regional planning in this country when they see us chucking it down the drain here,' said McAdden bitterly.

But the fear that the House was being asked to sign a blank cheque went further than that. Not only were they being asked to vote limitless funds; the amount of land to be developed was not limited either. There would be the airport and the seaport, and there would be attendant industry; but how much, no one knew. The Government had said there would be no primary industry established there, and they were grateful for that, but there was no way of knowing how much secondary industry there might be. 'There is no word in the Bill,' complained David Crouch, 'to say that there will not grow up on the shores of Essex and Kent an industrial complex which will be attracted by the biggest magnet ever created in the South-East – probably the largest seaport and airport in Britain, if not in Europe.'

Then there was the new town, accommodating perhaps a quarter of a million people; and the vast motorways – their route unknown, though the *Daily Telegraph* had regularly been publishing some scarifying forecasts. What would be left of south-east Essex when all this had been thrust upon it?

The Government had a majority of thirty-two on the amendment, with only two members, Robin Maxwell-Hyslop and Roger Moate, voting with the Opposition. But on the motion for the second reading, the Government majority fell to twenty-three. Maxwell-Hyslop and Moate were joined this time by McAdden, Braine and Hugh Fraser (Stafford) and at least seven Conservatives abstained.[3] On the Labour side, Hugh Jenkins voted with the Government on the second of these divisions and Michael Barnes abstained.

The debate completely failed to dispel the new sense of apprehension about the Maplin project, the feeling that once again a great technological programme was racing away, virtually out of control, with the politicians dragged along behind, clinging on for all they were worth. Unfortunately for the Government, the debate came just after a series of lost orders for Concorde; and there were several speakers eager to compare the two enterprises, as if to suggest that the eventual arrival of Concorde at Maplin, if it ever took place, would be a case of one white elephant alighting on another. The debate even produced, here and there, a distinct nostalgia for the old, simple days of Stansted. Douglas Jay was at the forefront, of course:

> Everybody who has examined the problem seriously, impartially and thoroughly [he said] knows that there is a far better case for enlarging Stansted than for this immense Maplin project ... When one allows for the motorways ... the disturbance caused by Maplin would injure far more people than another runway at Stansted, though they would be poorer people with less influence in the House of Lords.

But he was not alone. The continuing speculation about

whether the airport would ever be built served to postpone yet again that moment, once confidently expected, when the story would pass from the political stage and take its place as part of aviation history. The time had still not apparently arrived when politicians could lift their eyes from day to day developments and examine the whole tangled story to see what lessons, if any, it might produce. Such a course had been powerfully advocated by Peter Kirk, the leading Conservative campaigner against Stansted, in a letter to *The Times* on 5 March 1969 after the appearanec of Roskill's short list.

Having deplored the cost to so many local people of defending their rights against the threat from Whitehall, he went on to say:

> But what is surely more serious is the way in which the governmental machine worked – or failed to work. Here we had a case where two Ministers, in separate governments of opposite political persuasions, decided, on the advice of their departments, that Stansted was the right place for the third London airport. Yet when at last the problem is examined impartially, Stansted does not even pass the first test. There must be something wrong here.
>
> Neither Mr Julian Amery nor Mr Douglas Jay is a stupid man; on the contrary, they are both well known in political circles as extremely astute men, with formidable academic brains. How did they come to be so grossly misled?
>
> The need now must be for an urgent inquiry into the whole process of decision-making in government. Stansted is not the only example – even if it is the most flagrant – of the system going wrong, and we cannot possibly afford further errors on this scale. Surely it is now the duty of government, in the light of these events, to set up some kind of inquiry to see how decisions of this kind are made – and how they can be prevented in future.

In the light of all that has happened since, this plea seems even more worth making. So many years of wasted energy and wasted money have gone into the making of the decision on the third

London airport; not only on the part of Government but also on the part of those who have fought Government off at great personal expense – like Peter Kirk's constituents in Saffron Walden, finding £35,000 for themselves and a further £100,000 towards the expenses of the two county councils; or like the people in north Buckinghamshire of whom David Perman says, in *Cublington: Blueprint for Resistance*:[4]

> The people of Cublington paid thrice over to keep the airport from their door – first as taxpayers (the public cost of the Roskill Inquiry was £1,131,000), then as ratepayers (Bucks County Council spent £40,000 on its legal representation) and then as members of WARA. [WARA spent £43,713 on legal and technical representation.] If they also happened to be members of the National Farmers Union (which spent £6,000 on being represented before the Commission) they paid four times over.

They have, of course, the pleasure of knowing that posterity (or a majority of it) will applaud their fight, will conclude that they were right to resist and that Cublington was always the wrong site for a major airport. But many of them may regard that as a somewhat inadequate solace.

The one immediate moral of this story is that technology surged onwards and decision-making lagged behind. And that in itself suggests that the third London airport saga has a wider relevance; for the failure to anticipate the social, psychological and cultural effects of technological progress is one of the most persistent features of latter-day British government. Too often, it seems, we have proceeded by what might be called the 'big bang' theory of government, which means that it takes a large and dramatic event – a disaster perhaps – to alert us to what technology has dictated. People died in their thousands, year by year, from bronchitis and similar diseases, before the deaths in the London smog of 1952 made clean air a political issue. Supertankers chugged up and down our coast with their potentially abominable cargoes, even spilled many thousands of gallons – a bit at a time – into the sea before the striking of the *Torrey Canyon* upon the Sevenstones Reef made us see the threat they represented.

Town councils, confronted with a desperate need for new homes, built tower-flats, ignoring the warning sounds from the social service departments, only too glad to harness the new technology to a humanitarian cause; it was only with the collapse of Ronan Point that the fashion was killed, leaving behind it so many monuments to a policy which is now widely agreed to have been socially and psychologically disastrous. Concorde, a plane fathered by diplomacy as much as by aviation, was developed with totally inadequate thought for its environmental consequences – save that, in announcing it, Julian Amery, that fate-crossed Minister of Aviation, paused to note that the damage caused by sonic boom was expected to be 'negligible'. Only when people in the United States denounced it as unacceptable was it seen that unfortunate social consequences might turn out to mean financial catastrophe.

Motorways had been slung at ear-height past the windows of usually not very desirable residences for years before George Clark plastered the fate of Acklam Road over the national breakfast table. The laws of compensation were left unaltered while the scale of redevelopment and new development multiplied and multiplied again; the inspector at Gatwick inquiry found them inadequate, and Roskill, nearly twenty years later, made the same complaint. And it needed thalidomide to give us an official machinery to monitor new drugs.

That, perhaps, is the point at which the questions begin. It is easy to condemn the choice of Heathrow. The jet plane was well on the way; it should have been foreseen – but those who understood the implications were few, and other nations made the same mistake. It is still easier to condemn what happened over Gatwick – the breach of faith over Crawley (albeit attributable to the behaviour of the Russians, not after all the jolly comrades of wartime days); to condemn also the deception which led the

15

Government to indicate a minor airport before an inquiry and a 'main civil airport' once it was safely over.

But perhaps it is better to start with 1961, when, spurred by Robert Carr and his colleagues of the Select Committee, the Government set up the first Inter-Departmental Committee to look at the third London airport. Seen now, with hindsight, it looks as if a decision involving advanced technology was tackled with an attitude to what a new airport involved which was fatally imbued with the spirit of Gandar Dower and his odd job man taking the tickets at the gate. We were to build an airport; but for all the attention we gave to the environment and regional planning and to the wide social consequences of the project, we might have been building an airfield. So: the narrow departmental base. So: the concept of the Ministry responsible for planning as coming along behind, mopping up, dealing with the disruption of people's lives as if it were an epilogue, not a central part of the play.

These things were more widely understood two years later. Wilson by then had warned what aircraft noise could mean; 'intolerable' was his word, though it took two more years and a change of government before his message was accepted. He also warned what supersonic planes might do ('a much greater number of people would be affected by supersonic flight over land than are now affected by noise over airfields'). By the time of the Chelmsford inquiry (December 1965–January 1966), aircraft noise had become accepted as a fundamental issue, hotly contested by the two sides. But urbanisation, to which no official committee had yet drawn attention, was neglected by the Government (though not by the objectors).

Then, in 1966, the decision to suppress the Blake Report while an inter-departmental committee was given a chance to work it over and give vital government witnesses a second turn: that too argues a failure of foresight; a failure to see that even a good decision might be overthrown simply because of the suspicious way it had been handled, the lapse into secrecy, the flight from consultation. The failure to invoke the new econometric techniques in some other countries as a matter of course, and constantly commended by the Stansted objectors, argues lack of foresight too.

That leads inevitably to the troubles of 1967, when the BAA said that the need had now reached extreme urgency and the department – and the Government – agreed. It is hard to blame the BAA for making the most of their case; for it is easy to imagine the scandal there would have been had Heathrow and Gatwick jammed solid, like New York, and money-laden foreigners, in these days of balance of payments crisis, had found themselves forced to make do with countries better able to receive them. But why did the Board of Trade come up with the same answers? It does not seem to have been the case – as it was in the scandal over aviation contracts, when companies were held to have misled the Ministry of Aviation about their costs, and to have made profits which they later repaid – that Whitehall simply lacked the expertise to challenge the figures which the industry put before them.[5] Partly, perhaps, the difficulty was that the Board of Trade was in a way married to the BAA, in the way that government departments do become married to the interests with which their business is deeply intertwined.

Indeed, in a sense, the BAA *were* the Board of Trade, because several of the principal figures in the new authority had come over from Whitehall: George Hole, who had chaired the first Inter-Departmental Committee, was Chief Executive; Sir John Briscoe, their Director of Operations; Graham Hill, who had appeared before the Carr Committee at Stansted, who became their Director of Research and Projects; Norman Payne, who became their Director of Engineering; and Victor Harries, who became their secretary and solicitor. But it was not as if the arguments about urgency were never publicly disputed; there were even disbelievers within the Department. The significant thing was that the Minister was convinced, and carried his conviction in Cabinet.

Undoubtedly the failures in forecasting which occurred in the history of the third London airport did much to inspire misguided decisions by the politicians. Airport forecasting suffers from the same hazards as most other kinds of forecasting, from economic growth or unemployment to results in the four divisions of the Football League. What happens will depend on a variety of groups all along the line, and if you are wrong about any of them you are liable to be wrong with your final calculation.

The airport planner needs to make accurate assumptions about where potential customers live and where they will want to travel; he must also forecast the services which tour operators and airlines will decide to offer them, and the kind of aircraft in which they will be transported. That will also mean predicting the kind of aircraft which the manufacturers will be creating to cater for this traffic. The time span of these decisions is different in every case, and each decision in the chain is subject to sudden revision. The others in the chain may, indeed, revise their own assumptions because of what the airport planner himself is doing. Fears of airport congestion, for example – even the fear of being forced to fly out of Maplin – seemed at the start of the seventies to be accelerating the purchase of wide-bodied planes which would make it possible to carry more passengers without increasing the number of air transport movements. In Britain, as elsewhere, the forecasters had too often to work, if not in the dark, at least in a vague and misleading half-light. Their penetration of each other's information did not go deep enough. There was, perhaps, another failing too. Forecasting needs to be a continuous process, with today's figures treated as no more than the basis for tomorrow's updating. That makes for an uncertain kind of life: but perhaps it is the only way of dealing with a world as uncertain and fast-changing as that of modern technology.

Then there is the defence objection. The first and second Whitehall committees accepted it; yet to outsiders who strayed on to the scene – like Blake, like Brancker, like, eventually, Roskill – much of what was submitted by this ministry seemed to have the resounding ring of public nonsense about it. Shoeburyness was immovable in 1963 and again immovable in 1967, but seemingly almost irrelevant two years later. We shall never know, of course, since we are not to have an airport at Cublington (at least, not yet) whether the Ministry of Defence were as wrong about Brize Norton as Roskill supposed.

How far was this failure to penetrate their guard attributable to that mystique which goes with the word 'security'? That Whitehall mystique was comparable in status to that which used to hang round devaluation, until Anthony Barber finally dispersed it with a sentence or two slipped into a Budget speech – thus breaking a tradition, which, in Labour's day, had reached the stage where Ministers were solemnly adjured not even to mention the word.

And finally, Roskill. There are surely lessons here about the attempt to take decisions out of the political ring – only to have them forced back into it by the refusal of other people to respond in a non-political way. Politics defeated Roskill, just as, in local government reform, politics (again in the shape of Peter Walker) defeated Redcliffe-Maud; and just as politics, in the form of the Party and the TUC as well as the Cabinet and the Parliamentary Party, defeated *In Place of Strife*. Would there be pressure again, in such circumstances, to put one's faith in the impartiality of the judge, when the judge must blind his eyes to political necessity in a way that governments cannot?

There are some common strands in all this. Here was the biggest planning and investment decision of its kind that anyone concerned had ever grappled with. But the expert advice was inadequately subjected to challenge; the only challenge came from the protest movements, which were understandably held to be driven by self-interest, and which were then too hurriedly and shallowly written off.

The Minister relies on his Department. But if he relies on no one else, he may play a poor part in the departmental dialogue. That is the case made for the French system, where the Minister has a personal cabinet; good for stimulating challenge, but hard, perhaps, for the professionals to live with, knowing that those around the king may always be whispering against them. This course was advocated by the Labour Party before Fulton, but rejected. But a Minister can clearly gain from the presence of outside advisers, called in as Alan Day was by Crosland, to cast a new eye on entrenched assumptions – assumptions likely to be all the more entrenched when Ministers constantly come and go while the Department goes on for ever.[6]

It takes a bold Minister to pit his judgment against that of his official advisers. They have been at it longer, and can claim with certainty to have looked deeper. Richard Marsh, when Minister of Power in the Labour Government, complained to Cecil King (according to King) that he was expected to make decisions on important matters when his knowledge was only what could be put on the back of a cigarette card.

And Ministers may be as starved of time as they can be of information. Richard Crossman went to Harvard in April 1970 to deliver a series of lectures on government, characteristically discarding, in the interests of indiscretion, the approved texts he composed in London. (They are now contained in his book *Inside View; Three Lectures on Prime Ministerial Government.*[7]) His first confession was of sheer relief at escaping. 'For a week', he exclaimed joyously at the start of his first lecture, 'I shall be able to be completely detached. I haven't got any red boxes; I've got no documents. I haven't got the grind of the Ministry. I actually have time to think.' Crossman at that time was Secretary of State for Social Services, commanding a department which spent £6,000 million a year and touched the lives of almost everyone in the country.

But if the Minister fails to challenge the advice which comes before him, who else will do so? Back-benchers, however assiduous, are starved of resources; they have an improving research service in the library of the House of Commons, but nothing to match the research armoury of their counterparts in the United States.

The Opposition, in theory, is always snapping at the Government's heels, challenging its assumptions. But that cannot be relied on. The Shadow Minister is pitifully short of resources, compared with the man in occupation. The party organisations are not adequate to cope, though Conservative Central Office, as ever, is better equipped than Transport House.

There has, however, been a progressive use of policy groups, involving outside experts known to be sympathetic to the Party. Edward Heath, as leader of the Opposition, developed the network of policy groups which already served the Party and gave them a crucial place in building the platform for the 1970 general election. Between 1966 and 1970 the Party could call on twenty-nine groups with a combined membership of 191 MPs and peers, and 190 outsiders. In 1967 the Party set up a Public Sector Research Unit which commissioned studies on selected questions; those chosen included the Rotterdam Europoort and the third London airport. After the 1970 election, Labour instituted a network of policy groups on a much larger scale than before. According to the General-Secretary of the Party, Ron Hayward, who described them in a speech to a Fabian summer school in January 1973, there were forty-nine policy committees and study groups working out ideas 'on virtually every aspect of political policy'. They comprised over 250 people, including Members of Parliament, members of the Party's National Executive, academics, trade union officers, experts and rank and file members. Even so, it is doubtful if oppositions in this country will ever be able to mount a sustained and effective challenge to Government across the whole spread of policy while party politics in this country remain the shoestring affair they are today.[8] The policy groups, useful as they are, cannot always give the Shadow Minister what he most needs: a really well-researched answer to a specific question. For that reason, the sight of a Shadow Minister trying to grapple with policy can still be one of the saddest in politics.

Access to information is the key to the whole exercise. There is a great deal of essential information locked up on the Government side which it will only release if it chooses. Planning can be well advanced before anyone knows of it. The first Inter-Departmental Committee spent eighteen months working on a brief which was afterwards seen to have been inadequate. Had their lines of inquiry been better known, had more people been brought into the process of consultation, the objections which so impressed the Stansted inquiry might have been confronted earlier and a great deal of what happened afterwards might have been averted.

'One ought not to have a situation,' said the Treasury solicitor before the Franks Committee, 'in which the conflict of departmental policies is threshed out in public.'[9] The convenient way is for a solution simply to emerge; but it is not any likelier to be the right decision. The new super-ministries should prevent issues which before had to be resolved at Cabinet level from cluttering up the Cabinet agenda; but there is a compensating danger that they may now be resolved without challenge from outside the Department.

Inconvenient though it must have been to the departments concerned, the evidence suggests that outside intervention in airports policy has several times been beneficial. The Carr inquiry of 1960–61 exposed a reliance on assumptions rather than decisions; a state of affairs which was later to be blamed by Ministers for narrowing down the field of choice until an assumption became an inevitability. The Board of Trade feared the fate of Thales; they feared to move too soon in case their decision was out of date before it was executed; though that hardly squared with the arguments of urgency which were used by Ministers later on. The Blake inquiry – headed by a partner in a firm of estate agents – saw that the Stansted protestors had a more than parochial point. True, the evidence of their consultants had been compiled to a brief, that brief being to get the proposal abandoned; yet in compiling their case they had taken a more comprehensive and in the end more realistic view than the Government had done. It was Blake and Brancker too who challenged the sacrosanct nature of Shoeburyness.

Roskill at last provided the totally comprehensive view which the objectors had always wanted and which the size of the decision demanded – though the weapon the outsiders had long advocated, cost benefit analysis, turned out to be rather different in effect from what they had expected. The Roskill exercise, of course, was hopelessly beyond the resources of the Board of Trade. Yet that did not explain why even as late as 1966, in the review of the Peterson Committee, these techniques had not been attempted. The Board of Trade internal review did enough with them to foreshadow the eventual Roskill verdict on Stansted.

Of course there have since been improvements. Parliamentary scrutiny by select committee is becoming a thriving light industry, thanks to the Crossman reforms of the 1966 Labour Government and the further development which has taken place since the Conservatives came to power. Their cool impartial scrutiny is better adapted to the increasingly technical nature of political decision than the gladiatorial combat on the floor. Select committees have normally had more useful things to say than have come out of aviation debates.

The chamber is a good place for asserting truths but less good for honestly groping about in search of them. Reading aviation debates one often senses that they are happiest when they can settle down on traditional party lines, as when Mrs Tate hymned the joys of private enterprise and Bevan condemned its perpetual clamour for subsidy. But the temptation to look for party issues can be dangerous; the 'class war' view of Stansted helped to distort the Government's attitude to it. Yet it is in the chamber that the final confrontations occur. Successive select committees have hammered at the Concorde project and produced unanswerable criticisms of it, but they have not halted its progress. Debates on the floor of the House have suffered from the failure of the two sides to locate any clear-cut division between them (particularly, perhaps, because since 1970 Labour's chief spokesman has been an MP from Bristol, where Concorde keeps people in jobs).

The select committees may win the arguments but the power rests in the Government's hands. Only on the floor of the House (or in the behind-the-scenes manoeuvres which precede a likely showdown there), and there only very rarely, can a policy be broken. It was the knowledge that a vote on the floor of the House could not be carried which broke *In Place of Strife* and saved north Buckinghamshire from the third London airport. If the select committees are to expand, it may well be that the necessary saving of time can best be found by re-examining the work of standing committees, where it is difficult to find much productive return for the enormous time invested.

Edward Heath's 'think tank' (more formally, the Central Policy Review Staff) headed by the Labour peer, Lord Rothschild, which can seize on a great subject, possibly spanning ministries, and cast an extra-departmental eye upon it, might well have brought a more rational approach to the question of the third airport had it existed in time. There are, too, additional built-in inducements to administrative foresight now in the techniques of PAR and PESC (Programme Analysis Review and Public Expenditure Survey Committee).[10] What is more, consultation is better practised than it used to be. Labour under Harold Wilson introduced the system of Green Papers, enabling MPs and the world outside to kick about emerging policy before it sets into the semi-solid form of the White Paper. Clearly, any analysis of the kind advocated by Peter Kirk would have some progress to report.

But these, inevitably, are no more than tentative indications of the territory which may be worth exploring. It is no use confining yourself to turning over the old ground and prescribing cures for specific ills you may find there. The nature of the problem changes, and as fast as you mend one hole in the dam another appears. What is more, the technological terms of trade keep changing, almost too fast to measure. This year's best solution may be hopelessly anachronistic next year. 'Roskill,' said one close observer later, 'was working in 1969, but he was hopelessly imprisoned in the technology of 1967.' Or as James Barnes, Under-Secretary at the Department of Trade, told the Select Committee on Nationalised Industries – with what sounded li understandable weariness – in January 1970: 'the day will dawn when a decision will be taken, the airport will be constructed on a particular site, and I think it is very predictable that ten years after that, people will be wishing it was somewhere else.'

Despite that authoritative warning against wisdom after the event, it does seem fair to conclude that governmental attitudes to the demands of advancing technology, and governmental willingness to control it rather than to respond to it, were tested and found wanting in the search for a third airport. We shall not, if there is any mercy in the world, need to search for any more London airports. But there are other problems of equal majesty, not entirely beyond comparison with it, lurking elsewhere.[11]

Having subscribed to the concept of directing, or persuading, traffic into Foulness, the Government now had one more step to take if the conditions Buchanan attached to his recommendation were to be met; Foulness must be set in the context of a national airports policy.

Such sentiments were not calculated to enthuse the old Board of Trade hands now at the Department of Trade and Industry. They had been listening to arguments like this for a long time now and they did not see how it could be done. The fact was that the Board of Trade had been among those who had wanted Roskill (rather to his resentment[12]) to take on this task. As James Barnes told the Select Committee on Nationalised Industries on 26 January 1971:

I think everyone hoped two and a half years ago that the Roskill Commission would in effect plan a very great part of our aviation infra-structure, that is the South-East and the areas adjoining the South-East. But we do see today the kind of controversy that Report has engendered and I fear a wider plan covering the whole of the country would simply multiply that controversy by perhaps two or three times.

Down the road, however, there was another Department which didn't take quite such a gloomy view. The previous day, the *Daily Telegraph* reported that in a BBC interview, Walker, hymning the merits of his big new Department, had numbered among them its ability to work out, in conjunction with the Airports Authority, a national airports strategy. He didn't think the Airports Authority could do it alone: 'their interest is in what is most efficient for running an airport. There are other considerations apart from this.'

With the Maplin development, the men carrying the heavy piece of furniture seemed to have switched around. Now it was the Department of the Environment which was at the front, making the crucial decision, though in consultation at all times with its colleague at the back. One reason for this change was the wide-ranging nature of the Maplin development, in which the airport was only one part. But there did seem to be detectable in this new relationship, something of the go-getting image which Walker had stamped upon his Department.

The promise was duly fulfilled; the Civil Aviation Authority, set up in 1971 as a result of the findings of the Edwards Committee began commissioning studies from consultants as the basis for developing a national strategy. As it happens, the case for a national airports policy is probably weaker in the early seventies than it used to be. The time it was really needed, as I hope to have shown, was in the fifties. Yet the light which these two conversations, Barnes's with the Select Committee and Walker's with the BBC, throws on the attitudes of the two Departments does perhaps indicate – as did the change from Jay to Crosland – that the cast of mind of a Minister can make a difference, that Ministers are not the puppets of their civil servants that people sometimes cynically pretend.

The Maplin Development Bill survived its second hurdle – a Commons select committee, dealing with petitions from local interests, but debarred from examining the need for the airport (that was deemed to have been established by the vote on the second reading) – in March 1973. As the bill moved on to standing committee, the clamour to have the decision re-opened was growing. In April, the results appeared of the review within the Civil Aviation Authority of the previous forecasts of passenger demand and air transport movements. The report had a statesmanlike tone about it, as befitted the work of a group on which the Government – still staunchly defending the Maplin project – had direct representation. But even that did not disguise the fact that the review left the case for the airport looking less sturdy than before. Calculations within the CAA, based largely on a sophisticated and up-to-date analysis by the British Airports Authority and incorporating an understanding of airline intentions which was much fuller than that available to Roskill, suggested that there would be almost enough runway capacity at existing airports to cope until after 1985. It was capacity to handle passengers, not capacity to handle planes, which emerged as the most immediate problem: and one could hardly argue that one needed to build Maplin simply to deal with that.

It was still too soon to say, in the spring of 1973, whether it was the great jets or the dark-bellied Brent geese which would command the Maplin Sands a decade or so ahead. But the odds against the geese seemed to be shortening.

APPENDIX I – *Chronology of Third London Airport*

DATE	AVIATION MINISTER *Civil Aviation*	PLANNING MINISTER *Housing & Local Government*	EVENT	RELEVANT DOCUMENTS	TEXT REFERENCES
1952 July	A. Lennox-Boyd	H. Macmillan	Choice of Gatwick announced		44
1953 July	"	"	White Paper selects Gatwick	Cmd 8902	40, 44
October	*Transport and Civil Aviation* A. Lennox-Boyd	"	Merger of ministries of Civil Aviation and Transport		
1954 June October	" J. Boyd-Carpenter	D. Sandys	Report of Gatwick inquiry White Paper confirms choice of Gatwick	Cmd 9215 Cmd 9296	47 47–8
1957 April	H. Watkinson	H. Brooke	Report of Millbourn Committee urges study of third airport		68
1958	*Aviation* "	"	First jets at Heathrow		32
1959 October	D. Sandys	"	Ministry of Aviation created		

Date	Aviation Minister	Planning Minister	Event	Relevant Documents	Text References
1961	P. Thorneycroft	H. Brooke	Estimates Committee report on London's airports	HC 233 1960–61	63–73
November	,,	C. Hill	First Inter-Departmental Committee (Hole Committee) appointed		73–4
1963 March	J. Amery	Sir K. Joseph	Report of Wilson Committee on noise	Cmnd 2056	33–4
1964 March	,,	,,	Amery accepts IDC's Stansted recommendation	C.A.P. 199	76
July	,,	,,	Stansted plans shown at Harlow meeting		81
October	R. Jenkins	R. Crossman	Labour elected		
1965 June			BAA created		
December	R. Jenkins/F. Mulley	,,	Stansted inquiry (Blake inquiry) at Chelmsford		86–91
1966 March	F. Mulley	,,	Labour re-elected		
May	,,	,,	Blake report submitted		91
June	*Board of Trade* D. Jay	,,	Transfer of civil aviation responsibilities to Board of Trade announced by H. Wilson		
1967	D. Jay	A. Greenwood	White Paper backs choice of Stansted; Blake–Brancker reports also published	Cmnd 3259 Report of the Inquiry into Local Objections, MHLG (75–179)	94–102

APPENDIX 1 (cont.)

DATE	AVIATION MINISTER	PLANNING MINISTER	EVENT	RELEVANT DOCUMENTS	TEXT REFERENCES
June	D. Jay	A. Greenwood	Commons debate	HC Deb vol 749 no 221, cols 769 to 891	119–21
December	A. Crosland	,,	Lords debate	HL Deb vol 287 no 20, cols 861–1002	132–7
1968 February	,,	,,	Government announces decision to set up Commission	HC Deb vol 759 no 66, cols 667–674	140–1
			Report of Council on Tribunals.	Appendix to Council's annual report 1968.	137–40
May	,,	,,	Appointment and terms of reference of Roskill Commission	HC Deb vol 765 no 121, cols 32 to 40	141, 161, 167
July	,,	,,	First private meeting of Commission		
November	,,	,,	First public hearing	Roskill vols I and II	
December	,,	,,	Exeter College meeting		170
1969 March	,,	,,	Short list published		171
May–September	,,	*Local Government & Regional Planning* A. Crosland	Local hearings	Roskill vols III to VI	173–88
1970 April–August	R. Mason until June 1970, then *Trade and Industry* G. Rippon, J. Davies	*Housing and Local Government* P. Walker *Department of the Environment*	Final public hearings	Roskill Stage V	188
November	J. Davies	P. Walker	Buchanan says he cannot sign		189
December	,,	,,	Report sent to Secretary for Trade and Industry; summary published		189

Date 1971	Aviation Minister	Planning Minister	Event	Relevant Documents	Text References
January	J. Davies	P. Walker	Full Report published	Roskill Report	191
February	,,	,,	Lords debate	HL Deb vol 315 no 60, cols 812–918; vol 315 no 61, cols 928–1051	211
March	,,	,,	Commons debate	HC deb vol 812 no. 97, cols 1912 to 2046 and vol 812 no 98. cols 1911 to 2078	211
April	,,	,,	Choice of Foulness announced	HC Deb vol 816 no 128, cols 34 to 44	214

Postscript

During the summer of 1973, the battle over Maplin grew from a series of localised skirmishes into a continuing political confrontation. The opponents of the Maplin scheme, armed with the findings of the Civil Aviation Authority report, began to force the government back from one line of defence to another – and sometimes the new line looked woefully under-prepared.

Little was heard now about the need for runway or even for passenger capacity. The crucial element in the government's argument was the noise effect on existing airports if Maplin were not to be built. But on this point too the government's critics pressed them hard. Having got the traffic and passenger forecasts from the CAA, they challenged the government to produce noise forecasts to match; and when the government at first demurred, Tony Flowerdew, who had been deputy director of Roskill's research team, produced some of his own. In July, a Commons written answer disclosed that the Department of the Environment and the CAA were now at work producing the information which had been demanded.

The government were deeply suspicious of the anti-Maplin campaign. They suspected that the extremely skilful and well-orchestrated operation directed against them was the work of a group of interests, and especially the airline interests, which had never wanted Maplin from the start and were now willing to manipulate the evidence in the hope of frustrating the government's will.

That was the word dropped in the ear of potential rebels on the government's back benches: but it was not an interpretation of events which those Conservative MPs who distrusted the whole Maplin enterprise seemed eager to accept. The Essex and Kent MPs and the advocates of regional airports were now joined by others, notably the critics of the high rate of government expenditure, in outright opposition to the scheme. In June, the government were defeated in the Commons on an amendment, moved by Robert Adley, which wrote into the Maplin Development Bill a commitment designed to make the government consult with the CAA and Maplin Development Authority on all new technical evidence on the need for an airport, and to 'delay, vary, and desist from' its construction if that was what the evidence indicated. It fell well short of a death sentence on Maplin; indeed, some of the Conservative rebels thought the amendment should never have been tabled, calculating that it lessened the chance of defeating the government on the third reading. (The government were defeated by 17 votes on the amendment, with 15 Conservatives voting with the opposition; the third reading went through by a majority of nine.) But it was a deeply humiliating matter for such a commitment to be hung around the government's unwilling neck; and it erected a formidable new obstacle in the path of the Maplin project.

But one man, at least, was deaf to the clamour. Twice in the space of a week in mid-July, Edward Heath went out of his way to stake his own prestige on the success of the Maplin adventure – first, in a letter to a Conservative backbencher who wrote to him about aircraft noise at Heathrow; and then in his traditional end-of-term address to the 1922 Committee of Conservative backbenchers. Conservative Central Office then despatched a rallying call to its wavering Parliamentary troops: it did not carry the Prime Minister's name, but it had a distinct ring of Edward Heath about it:

> It is always easy to say 'no' to tomorrow: that is the cheap way out. It is always more difficult to say 'yes'; it will certainly demand some boldness, even some sacrifice. . . . A nation is judged not only for what it is now but for how it faces the future. And that demands self-confidence and resolution. Confidence in our national character; resolution to meet today the needs of tomorrow. . . .

But as the plans appeared, next day, for the new airport city, and for the motorway links which were to scythe their way through Essex, linking the new airport with a motorway ring, which there were, unfortunately, no plans yet to build, it was by no means clear that the spirit of adventure would be enough. A bad hole had been made in the government's case for an airport: the arguments of the 'do-without-it' school, which had seemed to have a strong tang of wishful thinking about them in the early days, now looked at least as substantial as anything produced in the airport's favour by the minister most directly responsible, Eldon Griffiths. As for the case for the seaport, that had never been made at all. It was, said Crosland, a classic case for inquiry by select committee.

The final confrontation seemed likely to come on an Order which the government was pledged to bring to Parliament before work on the airport could start. (The government had now, incidentally, ceased even to pretend that the airport would be ready in 1980: 'the early eighties' was Rippon's estimate in July). It had been the prospect of defeat on an Order which had broken Stansted: and it was distinctly possible that a similar fate would now befall Maplin too. The decision lay in the hands of Parliament, and most of all in the hands of the government's back-benches. Their weighing of the evidence, their balancing of their own innate doubts about the enterprise against the unflinching faith of their Party leader, would decide whether Maplin might in time become (in the immortal words of Eldon Griffiths) 'the world's first environmental airport'; or whether, instead, it would join Stansted and Cublington in the club of the great unbuilt.

Notes

CHAPTER I

1. Penguin, 1961.
2. See also page 195. Tony Aldous, in *Battle for the Environment* (Fontana, 1972), p. 249, quotes Buchanan as saying: 'I hate to say this, but I think I reached my conclusion *not on the evidence at all,* but on the basis of my experience in land use planning' (My italics).
3. Roskill Report, Note of Dissent, paragraph 32, p. 155. Buchanan's full case against cost benefit, argued on technical grounds as well as on grounds of innate suspicion, is in paragraphs 30–37.

CHAPTER 2

1. If you did, warned Willie Gallacher (Comm., Fife West), de Valera and his colleagues might come over the Border and take it off you.
2. Sir Peter Masefield, then head of the MCA Directorate of Long Term Planning, recalled later: 'Ted Heath used to go and fight on the committee and come back and cry on my shoulder about all the spokes put in the wheel by bumbledom.' Andrew Roth, *Heath and the Heathman* (Routledge and Kegan Paul, 1972).
3. See William Plowden, *The Motor Car and Politics 1896–1970* (Bodley Head, 1971) for a comprehensive documentation of governmental incoherence in face of the technological challenge of the car.
4. NNI – Noise and Number Index, the standard system of noise measurement since the report of the Wilson Committee.

CHAPTER 4

1. In April 1947, three months *before* the plan was announced, Sir Henry Self, Permanent Secretary at the Ministry of Civil Aviation, told a Commons Select Committee that the programme might have to be delayed because of the economic situation.

2. This was before the designation of a new town at Milton Keynes which greatly reduced the planning attractions.

3. See page 235.

4. Roskill Report, Note of Dissent, paragraph 38, p. 156.

5. In January 1973, the Secretary of State for the Environment, Geoffrey Rippon, rejected, on the advice of his Inspector, a £2·5 million development plan for Luton Airport. He said the airlines and holiday tour operators should recognise that the time had come to consider moving away from Luton or altering the pattern of their operations.

6. Paragraphs 40 and 41, pp. 156–7.

CHAPTER 5

1. It was actually 3·7 million.

2. See Appendix 1.

CHAPTER 6

1. Jenkins left Aviation for the Home Office on 23 December 1965 – seventeen days after the opening of the Chelmsford inquiry.

2. Acceptance of Ministry of Defence assertions by other departments was also the subject of outside criticism. In the *Guardian*, David Fairhall, Air and Defence correspondent, regretted that this evidence had not been more vigorously challenged at Chelmsford. A subsequent *Guardian* leader condemned 'the placid acceptance that the country's biggest transport decision for decades should be subservient to the squatters' rights claimed for the Thames-side artillery ranges'.

3. See p. 87 above.

4. See p. 55 above.

5. He succeeded R. A. Butler at the by-election of 23 March 1965.

6. Griffiths was later to become Department of Environment Minister with direct responsibility for the Maplin project.

7. See p. 137.

8. Weidenfeld and Nicolson/Michael Joseph, 1971.

9. These figures were given in the Lords debate in December. They were a revision of previous estimates.

10. This was the first PLP meeting at which an official briefing was later given to the Press.

11. See p. 55.

12. *The Times* reported that ten Labour MPs seemed to have abstained including Newens, Mrs Short, John Parker (Dagenham), Charles Mapp (Oldham East) and Arthur Lewis (West Ham North). Newens and Lewis were present for a division later that evening as was Peter Jackson (High Peak) who also failed to vote with his Party on Stansted.

13. Wilson says in *The Labour Government 1964–70: a Personal Record* that Jay was dropped on grounds of age, not anti-Marketeering. That was not the interpretation of events held by anti-Marketeers.

CHAPTER 7

1. Cape, 1956.

2. Penguin, 1971.

3. Paragraph 40. See p. 95.

4. Cape, 1970.

5. It wasn't.

6. See R. E. Wraith and G. B. Lamb, *Public Inquiries As An Instrument Of Government* (Allen and Unwin, 1971) on which this account is based.

7. It was later challenged by the Ministry of Housing and Local Government (*Daily Telegraph*, 24 February 1968).

CHAPTER 8

1. Arlington Books, 1971.

2. Another valuable ally was the Assessor of the Chelmsford inquiry, J. W. S. Brancker, who became a consultant to TEDCO.

3. The exclusion of TEDCO by Peter Walker, though not irrevocable, led to some sense of disillusion in the town. At a meeting at Southend on 22 March 1972, Norman Harris and the eminent planner Lord Holford, a TAG consultant, sympathised

with each other over their disappointment, Harris remarking: 'We have both lost all our money.'

4. See *The Role of Commissions in Policy-Making* edited by Richard A. Chapman (Allen and Unwin, 1972), especially pp. 174–5. Chapman says there should be 'urgent research' within the Civil Service Department on the question of designation.

5. In *The Times*, 20 December 1972.

6. 'The kind of membership I have in mind would include perhaps a traffic engineer, an aviation expert, an economist, a businessman and a regional planner, in addition to a planning inspector' (20 May 1968).

7. See Plowden, op. cit., pp. 328–9.

8. And was indeed forecast by Harford Thomas in the *Guardian* in December 1967, two months before the Commission was announced. Examining the choice in the light of the principles enunciated in *Traffic in Towns*, he found that Foulness was the unmistakable indication.

9. In a symposium on 'Planning by Inquiry' organised by the Royal Institute of British Architects and the Town and Country Planning Association, 26 April 1971. See Aldous pages 246ff.

10. Fontana, 1972.

11. For the extreme condemnation of CBA, see Peter Self in the *Political Quarterly* vol. 41 no. 3. Also Self's exchanges with Peter Hall in *New Society*, 28 January and 4 February 1971. For a milder critique, see Malcolm MacEwen, *New Statesman*, 6 March 1970.

12. See evidence of Stephen Hastings, MP to Roskill (Document 6019).

13. See p. 141.

14. Mr Hooley (Lab., Sheffield Heeley): Will the terms of reference entirely exclude the possibility of siting the airport outside the South-East, since it could gravely aggravate the congestion which it is designed to relieve?

Mr Crosland: No sir, they do not entirely exclude this. It would be open to the Commission to say that there was no need for a third airport or that it should be indefinitely postponed. This is unlikely in practice, because almost everyone who has discussed this matter is convinced that some time in the middle seventies there will be a need for a third London airport.

15. Though his figures were largely taken from the East Anglia region, and Thurleigh is outside it.

16. Lawn had campaigned against Stansted as a member of Harlow Labour Party. He did not in the event fight Huntingdon in the 1970 election. Curran himself was the candidate.

17. For a comprehensive account of the Cublington campaign, see David Perman, *Cublington: A Blueprint for Resistance* (Bodley Head, 1973) which the author very kindly allowed me to see before publication.

18. See Roskill Report, 6.74, p. 54, and Commission's Papers and Proceedings, vol. VIII, part 2, section 4.

19. Evelyn de Rothschild. WARA numbered among its vice-presidents four Conservative MPs, James Allason (Hemel Hempstead), Stephen Hastings (Mid-Beds.), Sir Spencer Summers (Aylesbury) and Arthur Jones (South Northants); and two Labour MPs, Robert Maxwell (Buckingham) and Gwilym Roberts (South Beds.).

20. Volume V, Stage 2 local hearings, Cublington (Wing). See also Perman, *passim*.

21. Part of this case, though regrettably by no means all, is reported in *The Best of Beachcomber*, selected and introduced by Michael Frayn (Penguin, 1966).

22. See Perman, pp. 156–163.

23. Derrick Wood returned the remarkable figures of 6.4– 6–1–1.

CHAPTER 9.

1. Note of Dissent, paragraph 56, p. 159.

2. See p. 17.

3. 'Small' however was a curious description of Luton, which in 1970 handled nearly two million passengers (the third highest figure for any British airport). By 1971 this figure was fast approaching three million, putting Luton roughly on a par with

the national airport at Brussels and ahead of Le Bourget. See the excellent statistical returns in the BAA annual reports, especially pp. 76–7 of 1971–2.

4. See Chapman, op. cit. This book, produced by sociologists at provincial universities, exhibits a great distrust of what goes on at Oxford. However there was quite a widespread assumption that the Donovan Commission had been fatally nobbled by Nuffield.

5. See Aldous, op. cit., chapters 1 and 2, especially p. 35.

6. Penguin, 1969.

7. See the Commons debate on Maplin, 8 February 1973.

8. Professor Wibberley. See Roskill Report, 6.58.

9. See Geoffrey Sheridan in *Campaign*, 30 April 1971.

10. Written reply by Michael Noble, Minister for Trade, 27 July 1971.

CHAPTER 10

1. See p. 168.

2. The name Maplin had understandably been preferred to Foulness. This could reasonably be justified in that the airport would be on the sands (Maplin) rather than the island (Foulness).

3. Conservative MPs known to have abstained were Brian Harrison (Maldon), Robert McCrindle (Billericay) and Norman Tebbit (Epping – a former airline pilot) from Essex; David Crouch (Canterbury) from Kent; and Robert Adley (Bristol NE), John Tilney (Liverpool, Wavertree – a champion of the Liverpool airport at Speke) and Richard Body (Holland with Boston) from further afield. Some other potential abstainers seem to have been deterred by the fact that the Maplin debate took place at the height of the controversy over the Counter-Inflation Bill. One senior Conservative MP, a persistent critic of high government spending, told me afterwards that he believed Maplin should be the first casualty on any list of Government spending cuts, but he had not withheld his vote, 'because you can't get into too many rebellions at the same time'.

4. Bodley Head, 1973.

5. See the Report of the Lang Committee into the pricing of Ministry of Aviation contracts, July 1964.

6. See Appendix 1.

7. Jonathan Cape, 1972.

8. For proposals for financing political parties see Dick Leonard, MP in the *Observer*, 13 May 1973. Also reply by Douglas Hurd, in the *Observer*, 15 July 1973.

9. Quoted in Ian Gilmour, *The Body Politic* (Hutchinson, 1969).

10. See Sir Richard Clarke, *New Trends in Government* (HMSO 1971).

11. On all these points, see Edmund Dell, *Political Responsibility and Industry* (Allen and Unwin, forthcoming).

12. Roskill Report 2.12, pp. 8 and 9.

Index

Air Ministry, 24–7, 30, 32, 39, 40, 42, 63, 72 (*see also* Ministries of Civil Aviation, Transport and Civil Aviation, Aviation, Board of Trade)

Air Navigation Acts, 23, 32, 33

Amery, Julian, 73, 76, 79, 120, 223, 225

Aviation, Ministry of, 34, 47, 65–71, 72, 74, 86, 89–91, 100, 143, 227 (*see also* Air Ministry, Ministries of Civil Aviation and of Aviation, and Board of Trade)

Barnes, James, 53, 54, 234–6

Bedfordshire Airport Resistance Association (B.A.R.A.), 176–8, 207

Bedfordshire County Council, 176, 182, 212

Benn, A. N. W., 233

Beswick, Lord, 95, 97, 145, 218

Betjeman, John (now Sir John), 17–19, 147, 180, 205

Birmingham Airport (Elmdon), 52–4, 61, 192

Bishop's Stortford, Herts., 43, 44, 78, 81, 83, 84, 91, 139, 144

Blackbushe, 39, 40, 43, 44, 61, 62, 64, 65, 68

Blake, G. D., inspector at Stansted Inquiry, 87–97, 99–101, 106, 145, 175, 197, 226, 228, 232 (*see also* Stansted: public inquiry)

Board of Trade, 53, 54, 56–8, 95, 107, 114, 117, 122, 125, 128, 130, 132, 134, 143, 145, 146, 162, 163, 166, 172, 200, 227, 232, 233, 235 (*see also* Air Ministry, Ministries of Civil Aviation, Transport and Civil Aviation, and Aviation)

Braine, Sir Bernard, 209, 220, 222

Brancker, J. W. S., assessor at Stansted Inquiry, 87, 92–4, 96, 99, 100, 105, 145, 175, 228, 232, 247 (*see also* Stansted: public inquiry)

British Airports Authority (B.A.A.), 28, 52, 53, 57, 101, 102, 112, 117, 132, 160, 168, 174, 191, 195, 214, 215, 227, 236, 237

British Broadcasting Corporation (B.B.C.), 30, 82, 203, 206, 236

British European Airways (B.E.A.), 58, 65, 195

British Overseas Airways Corporation (B.O.A.C.), 195

Brize Norton, Oxon, 175, 182, 229

Buchanan, Colin (now Sir Colin), chapters 1 and 9 *passim*; 14, 17, 161–, 181,3 189, 245

dissents from Roskill, 14, 17, 18, 19, 166, 170, 171, 189, 195–200, 204, 205, 211

on cost-benefit analysis, 17, 162

on national airports policy, 56, 57, 199, 235

on regional airports, 60

Buckinghamshire County Council, 166, 176, 182, 212, 224

Butler, R. A. (now Lord Butler), 64, 75, 76, 79, 86, 246

Campbell, Sir Colin, inspector at Gatwick Inquiry, 45, 93 (*see also* Gatwick: public inquiry)

Carr, Robert, 63, 66–73, 119, 120, 226, 232

Chalk pit case, 103, 137, 138

Civil Aviation Acts, 33, 52

Civil Aviation Authority (C.A.A.), 220, 221, 236, 237

Concorde, 221, 222, 225, 233

Coventry Airport, 53, 54, 57

Crawley, Sussex, 43, 44, 46, 47, 182, 225

Crosland, C. A. R., 19, 110, 114, 123–32, 135, 138, 140, 144, 145, 150, 161, 162, 166–9, 171, 197, 198, 203, 218, 220, 230, 236, 248, 249

Crossman, R. H. S., 132, 140, 142, 158, 160, 230, 233

Crouch, David, 210, 220, 221, 250

Croydon, 24–6, 30, 39, 43, 44, 46

Cublington (Wing), Bucks., chapters 8 and 9 *passim*; 49, 59, 60, 180, 229

on Roskill short list, 15, 170–2

local hearings (Aylesbury), 178, 179, 182–7

effects of airport on locality, 35–7

campaign to prevent airport at, 17, 18, 165, 166, 181–8, 199–214, 224

selected by Roskill Commission, 13–17, 165, 190–5

choice opposed by Buchanan, 15–19, 195–9

rejected by government, 19, 199–215

debated in Lords, 211; and Commons, 168, 211

Commons early day motion on, 209

implications for regional policy, 60, 61, 197, 198

Ministry of Defence view, 175